BUILDING CONSTRUCTION

Site and Below-Grade Systems

BUILDING CONSTRUCTION

Site and Below-Grade Systems

James Ambrose
University of Southern California

VNR VAN NOSTRAND REINHOLD
New York

Copyright © 1991 by Van Nostrand Reinhold

Library of Congress Catalog Card Number 91-15800
ISBN 0-442-00293-9

Manufactured in the United States of America

Published by Van Nostrand Reinhold
115 Fifth Avenue
New York, New York 10003

Chapman and Hall
2-6 Boundary Row
London, SE1 8HN, England

Thomas Nelson Australia
102 Dodds Street
South Melbourne 3205
Victoria, Australia

Nelson Canada
1120 Birchmount Road
Scarborough, Ontario M1K 5G4, Canada

16 15 14 13 12 11 10 9 8 7 6 5 4 3 2 1

Library of Congress Cataloging-in-Publication Data

Ambrose, James E.
    Building construction : site and below-grade systems / James Ambrose.
        p.   cm.
    Includes index.
    ISBN 0-442-00293-9
    1. Foundations.   2. Building sites.   I. Title.
TH2101.A57   1991
    690'.11—dc20   91-15800
                                                                        CIP

Preface

This book treats a subject that is of interest to a wide range of people with various relationships to the designing and constructing of buildings. There is at any given time a great mass of information on these topics which may be accessed for application to the many tasks of designers, builders, suppliers, and others. The enormity of this information resource is at once reassuring to those who regularly encounter needs for it, and overwhelming to just about everyone who needs to figure out how to use it.

This book relates the topic of building construction to some of the basic problems of building and site design. The basic assumption here is that the need to design precedes the need to build, and that real concern for how to build comes from a desire to build something. This is the normal process of development for designers and others who start from the point of desiring a building and then proceed to determine what it should be. Intense concern for specific consideration of building materials, systems, and details of construction thus emerges at a later stage of design, typically after the general form, size, and essential nature of the building are already proposed.

A major problem with using the great mass of available information about building construction is that it is largely not oriented to the purposes of education or design; thus it is not user-friendly to the student or apprentice in architecture, or to any others who do not already have a broad grasp of what buildings and building construction are all about. Rich as they may be as information sources, *Sweets Catalog Files*, the *CSI Spec-Data System*, and *Architectural Graphic Standards* are not friendly to inexperienced users. You have to already know the trade lingo, what you are looking for, and pretty much why you are looking for it to make effective use of these resources.

The latest, most extensive, most detailed information about building construction is largely produced by persons engaged in the manufacturing, supplying, specifying, purchasing, regulating, financing, or insuring of building materials and products, and in their applications to making buildings. The information presented is often slanted toward the specific concerns of those who produce it. Thus it is to be expected that materials forthcoming from manufacturers and suppliers are shaped to the purpose of promoting sales of their products, that insurers and financiers have strong economic concerns, and that specification writers and enforcers of building codes are somewhat paranoid about specific language and terminology that is legally binding.

This book attempts to be user-friendly to the person who is basically more interested in buildings in general and somewhat less in the specific concerns for their construction. The basic purpose here is to develop a general view of buildings and the problems of their design as they relate to the eventual need to construct them. Highly specific, detailed information, such as how to attach gypsum drywall to wood studs, can be pursued through appropriate sources once the real need for the information is established. The reasons for having the drywall, for attaching it to studs, for having wood studs—and indeed, for having the wall in the first place—need to precede the search for the specific information.

Buildings are complex and the topic of their construction is correspondingly extensive. In order to cut down the size of this presentation, the topic here is limited to concerns for the development of the construction for elements that are below grade level or are parts of the site development: primarily basement walls and floors, foundations, site pavements and retaining structures, and other building-related site elements. That leaves a lot of the building not treated. This book is, in fact, the third volume of a series. The first volume treated the subject of the building's basic shell that forms the enclosure: primarily roofs, exterior walls, windows and doors. The second volume treated the subject of the building's interior: primarily floors, interior walls, ceilings, and stairs. A subsequent volume is to consider construction related to the development of building service systems.

My views of the need for this book and my ideas for its contents have emerged from many years of experience as a building designer and a teacher. I am grateful to all my former students, coworkers, and others whose feedback of frustrations have helped shape my views and ideas. I am also grateful to the many people at Van Nostrand Reinhold who have helped to bring my rough materials into being as an actual book; particularly my editors Everett Smethurst and Wendy Lochner as well as Cindy Zigmund and Alberta Gordon.

I am also grateful to various organizations who have permitted the use of materials from their publications, as noted throughout the book.

Finally, I am grateful to the members of my family who steadfastly support and tolerate me in my home office working situation. For this book, I am particularly grateful for the assistance of my wife, Peggy, and my son, Jeff.

James Ambrose

Contents

1

Introduction

The design of site and below-grade construction is a major area of concern in the general design of buildings. This chapter presents discussions of the issues and specific problems of site and below-grade construction and how the topic is developed in this book.

1.1 GENERAL CONSIDERATIONS

This book deals with construction on and in the ground, as related to a building and its site. In its broadest sense, this topic has many aspects, ordinarily involving the concerns of many designers. The general form, functional problems, and appearance of building sites are problems of major concern to architects, but also to planners, landscape architects, civil engineers, and many other persons.

On and below the ground surface of a site are many items of concern to the site development, the building, and various building utilities and services. Many individual designers must deal with these items, but the whole effort must be carefully coordinated if chaos is to be avoided. This is further complicated by the fact that the first work on the site—often preceding real completion of all the design work—is the regrading and excavation necessary for the beginning of construction.

While appearance and general spatial functioning of sites are critical concerns, there are many other design considerations for sites, including:

Surface drainage and disposal of runoff water.

Development of plantings.

Placement of the building on the site.

Development of building foundations and below-grade construction.

Hookup to off-site services.

Site edge development regarding existing streets and properties.

Site lighting, both natural and artificial.

Sound, for privacy and general noise control.

Air quality, as affected by exhaust air.

Fire safety, in terms of exiting of building occupants and access by fire fighters.

Security for people and contents of the building.

1

The shaping and general constructing of building sites must respond to all of these concerns—and, in many situations, to various additional ones.

The primary topic of this book is not site development in general, but rather the basic construction associated with it. However, the selection of materials, components, systems, and details for site construction impinges on the work of the designers of all of the aspects of the building and site. While it is possible to consider only a few concerns, or even a single one, all of the potential effects of choices for the construction must eventually be dealt with.

While the building site is a definable place and a specific design concern, it must also relate to the general building enclosure, the building structure, access to the building, and the many building service systems. In the end, the whole building and its site must be designed, even though specific situations may be dealt with as individual problems.

The point of view here is that of the designer and what is of prime concern in the design process. The writing of specifications, management of construction, and production of materials and components for construction are necessary activities that will be considered, but design is the focus.

1.2 BUILDING CONSTRUCTION AS A DESIGN PROBLEM

The making of buildings is a big business, affecting many people. However, the development of the materials in the book is based, in part, on the attitude that the determination of building construction is a design problem. The solving of problems, large and small, with regard to the construction work affects the design—or at least it ought to. If certain details or dimensions are not feasible to achieve with a particular material or system, designers should accept the facts, or prove by some means that what is proposed for the design can truly be accomplished. Design work must be informed, not necessarily in early sketching stages, but as soon as any serious commitments are made and the detailed development of the work proceeds.

It is possible, of course, to make the construction itself a major design determinant. However, respect for correctness and expression of the construction is one thing; reverence for it is another. Many architects are more concerned with form, space, appearance, illusion, metaphor, symbolic relations, human response, or other matters, and care somewhat less about the expression of the construction or the revealing of the building's functioning elements.

For most designers the actual work of design usually is a very personal matter. Launching points, decision sequences, and the relative weights of different values vary considerably from designer to designer as well as from project to project with a single designer. Approaching the subject of building construction as an architectural design issue simply means that concerns for the construction emerge as necessities in the process of design. The designer does not set out to produce some construction, but rather encounters construction considerations in the normal process of designing a building. First comes the concept of the building and all of its planning concerns— then the problems of making it.

There is no intention to advocate any particular design philosophy in this book. Proper construction is considered essentially as a simple, pragmatic matter. It is considered to have a reasonable value and to be something that deserves serious attention, but it is not essential that it command dominance in all design decisions.

1.3 HOW TO USE THIS BOOK

The basic purpose of this book is to help the inexperienced designer, whether in a school design class or in an apprentice position in a design office. However, anyone who is interested in the topic of building construction and its effects on building design should be able to derive some use from this book.

There is no single method for the most effective use of the materials in this book. Readers are sure to vary considerably with regard to interests, needs, and backgrounds. While the materials have been arranged with some logical sequence in mind, it is anticipated that the typical reader will take up the materials in a random fashion. This is to be expected all the more if the reader is seeking some help in the progress of a design work.

We do not mean to discourage any reader from taking up the book materials in the sequence presented. This is likely to make most sense if the reader is essentially uninformed in the general topic. Development of a fluent construction vocabulary is also to be generally encouraged, as the discipline of proper terminology is essential for accurate communication in the building design and construction business world.

Finally, for many readers, this book may serve a principal function in pointing out sources for additional information. General major sources are discussed in the next section, and various special sources are cited throughout the book in the discussions of specific topics.

1.4 SOURCES OF DESIGN INFORMATION

Information about building construction is forthcoming from many sources and relates to many potential uses (see Figure 1.1). For the specific purpose of supporting design activity, some particular sources and applications are of special value. This is a highly variable situation, responding to many considerations for a particular design activity. Significant variables include the following:

Stage of the design process—from very preliminary to final construction directives

Nature of the design project—large or small, luxury or low-cost, specific constraints (codes, special occupant problems, etc.)

Working context of the designer—time constraints, design budget, size of the design team, access to information sources

Individual types of information and types of information sources may be of particular value, depending on the design work as qualified by the preceding, as well as by many additional considerations. Some major categories of information sources that are generally supportive of design work are the following:

Books—including ones of broad scope, such as general references on building construction, and ones of highly specialized nature with important data for a very specific problem

Industry-supplied materials—mostly supplied for promotional purposes, but nevertheless usually indispensable for specific product data

Records of designs—statistical, graphic, or photographic records of proposed or actual construction

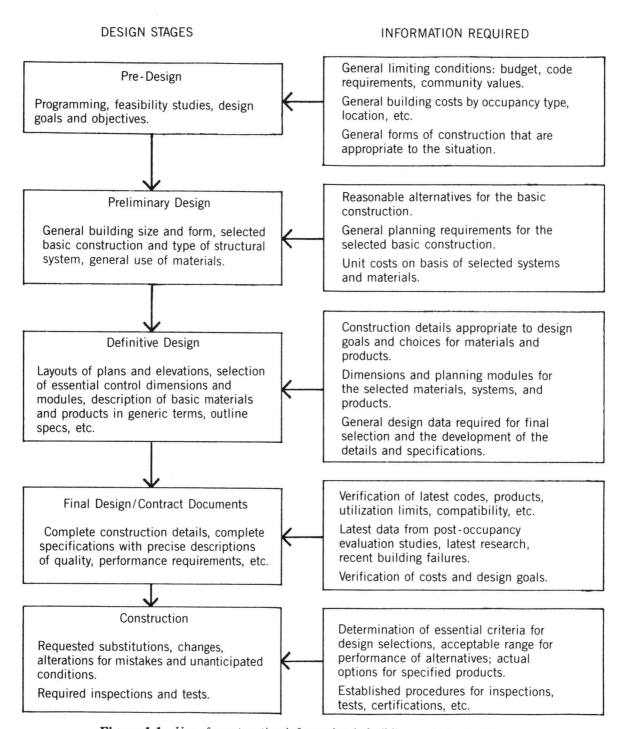

DESIGN STAGES   INFORMATION REQUIRED

Pre-Design

Programming, feasibility studies, design goals and objectives.

General limiting conditions: budget, code requirements, community values.

General building costs by occupancy type, location, etc.

General forms of construction that are appropriate to the situation.

Preliminary Design

General building size and form, selected basic construction and type of structural system, general use of materials.

Reasonable alternatives for the basic construction.

General planning requirements for the selected basic construction.

Unit costs on basis of selected systems and materials.

Definitive Design

Layouts of plans and elevations, selection of essential control dimensions and modules, description of basic materials and products in generic terms, outline specs, etc.

Construction details appropriate to design goals and choices for materials and products.

Dimensions and planning modules for the selected materials, systems, and products.

General design data required for final selection and the development of the details and specifications.

Final Design/Contract Documents

Complete construction details, complete specifications with precise descriptions of quality, performance requirements, etc.

Verification of latest codes, products, utilization limits, compatibility, etc.

Latest data from post-occupancy evaluation studies, latest research, recent building failures.

Verification of costs and design goals.

Construction

Requested substitutions, changes, alterations for mistakes and unanticipated conditions.

Required inspections and tests.

Determination of essential criteria for design selections, acceptable range for performance of alternatives; actual options for specified products.

Established procedures for inspections, tests, certifications, etc.

Figure 1.1   Use of construction information in building and site design.

For any specific design work, the truly usable or necessary references may relate as much to the particular needs of the individual designer as to the specific nature of the design activity. The education, training, developed skills, and previous experience of the designer may well make certain general references indispensable or generally require help only for very special situations.

This book is not intended as an inexhaustible source of specific information about all of building construction. Although the view is broad, the scope of the presentations here is narrowly bound by specific concentration on the needs of the designer, the general nature of design work, and particular interest in building sites and below-grade construction. It is generally assumed that the reader, whether in a school or in a design office, has access to some general information sources. These include the following basic items:

One or more general texts on basic building construction materials, products, and processes

One or more general references on standard details of building construction

Some industry-supplied information; as generally represented by *Sweets Catalog Files* or a selected, personally assembled set of individual entries as obtainable from individual manufacturers or agencies

It is assumed that perceptive readers are aware of some of the pitfalls in using information sources. We refer particularly to the major concerns for timeliness, neutrality of the suppliers, and feasible applicability to the design work at hand. In the last regard are such concerns as local codes, availability of products and services, and climate differences.

Building construction is largely achieved with commercially developed products. The suppliers of these products do a lot of hard selling to compete in the marketplace, and information obtained from them is quite understandably not of a neutral, objective nature. You can't expect to be able to judge the relative appropriateness of a product to a specific application on the basis of the information supplied by the product's supplier.

Many references have been used in the development of the work presented in this book. Many of these sources are cited for very specific information, but the general development of the work derives often from a collective consideration of many sources. For general reference, we acknowledge—and recommend to the reader—the following sources:

1. *Architectural Graphic Standards,* 8th ed., Ramsey and Sleeper, New York: Wiley, 1988.
2. *Fundamentals of Building Construction: Materials and Methods,* 2nd ed., Edward Allen, New York: Wiley, 1990.
3. *Construction Materials and Processes,* 3rd ed., Don Watson, New York: McGraw-Hill, 1986.
4. *Construction Principles, Materials, and Methods,* 5th ed., Olin, Schmidt, and Lewis, New York: Van Nostrand Reinhold, 1983.
5. *Sweet's Catalog Files: Products for General Building and Renovation,* New York: McGraw-Hill, annually published.

6. *Mechanical and Electrical Equipment for Buildings,* 7th ed., Stein, Reynolds, and McGuinness, New York: Wiley, 1986.

For specific problems and applications, many additional references are cited throughout the book.

2

Architectural Components

This chapter presents materials relating to the most ordinary and generally indispensable parts that occur in the majority of buildings and with some of the design concerns that arise in producing them. While designers are necessarily concerned with all of a building's parts, the concentration here is on those parts most directly involved in achieving the building's site, foundations, and general subgrade base, which is the central topic of this book.

Design of any selected building part must be done in the context of the problems of the whole building. Dealing with the individual elements of building construction requires some attention to their eventual coordination in the whole constructed building. Putting the parts together is the theme of the work in Chapter 7 of this book, where the whole construction of several buildings is presented.

The objective in this chapter is to deal generally with the various concerns for the basic building component parts. For real situations, consideration must be given to the means for producing the parts in terms of available materials and products. Chapter 3 discusses basic construction materials. Chapter 4 covers the various products used to achieve basic building components. Chapter 5 presents some special, although highly crucial, concerns. Chapter 6 considers the general problems of systems.

2.1 INDIVIDUAL COMPONENTS

Figure 2.1 shows some of the basic components that may constitute a building's subgrade base and site development. These and other fundamental site elements are discussed in the following sections in this chapter.

Actually, the major "component" that must be dealt with is the site. The site form, soils, and various existing features constitute a given resource as well as typically some problems that require design solutions. Placing the building on the site and generally integrating all the construction elements in and on the ground is a major design management problem—what we will generally refer to as the full development of the site.

2.2 BUILDING FOUNDATIONS

Foundations essentially consist of transitional elements between the building's aboveground superstructure and the ground. Basic problems involve the resistance

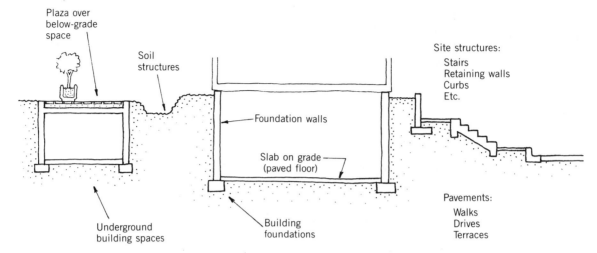

Figure 2.1   Basic construction components.

of vertical gravity loads, resistance of lateral load effects (wind and earthquakes), and generally providing a base for construction of the building.

The three basic types of foundation elements are those shown in Figure 2.2. The simplest and most common foundations consist of so-called shallow bearing foundations, also called footings. These typically consist of bearing pads of reinforced concrete of sufficient plan size to resolve vertical forces into a low value of bearing pressure on the soil.

When soils of sufficient bearing capacity do not exist at the location of the bottom of the building—or when the loads are simply too great for bearing pressure resolution—special foundations, called deep foundations, must be used. One form is the pile, a shaftlike element that is driven into the ground like a large nail. Timber poles (tree trunks) were used for this in ancient times and still are. More often, however, other forms are now used, as discussed in Section 4.5. Accuracy of placement of piles is quite limited, so piles are ordinarily used in groups.

The other form of deep foundation consists of an excavated shaft that is advanced

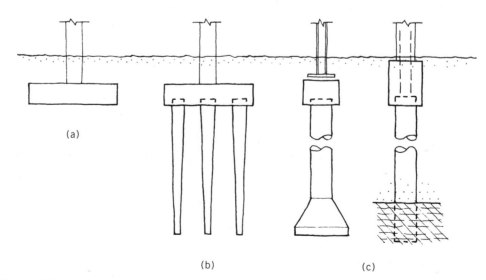

Figure 2.2   Basic types of building foundations: (a) Shallow bearing footings. (b) Driven piles. (c) Excavated piers.

down some distance and then filled with concrete to form a stiltlike column under the building. This is called a pier or caisson.

Both piles and piers may be seated in soil or advanced to bear on rock. When the excavated shaft bears on soil, it is common to splay or spread its bottom, forming a bell-shaped, truncated cone at the bottom of the round shaft, as shown in Figure 2.2(c).

These basic elements—footings, piles, and piers—are typically used with various other elements to form a complete building subgrade base. This may include grade beams, foundation walls, ties for isolated foundations, pile caps, and so on. Exact requirements will depend on the building size and form, the type of construction, the general type of foundation system, and various possible soil conditions. Several different situations are illustrated in the case studies in Chapter 7.

2.3 BASEMENTS

Many buildings consist essentially only of abovegrade construction and spaces, with a minimal ground penetration made basically to achieve the foundations and a base for construction. However, if interior spaces are extended below the ground level, the below-grade portion is called a basement.

Basements that exist only beneath the overhead building consist primarily of enclosing walls and floors. Essential problems, as shown in Figure 2.3, consist of retaining the soil outside the walls, preventing intrusion of water, and handling various factors for comfort of occupants when the spaces are used other than simply for equipment or storage.

When basements exist, they are usually points of entry and exit for various building services that are delivered underground—typically water, gas, and sewers, but also possibly underground power and telephone lines.

Basement walls also frequently serve as bearing walls or spanning grade beams in conjunction with the building foundation system. Architectural planning and general detailing of the construction must deal with all of these relationships.

When basements are deep, extending considerably below finished ground level, a major consideration is often the achieving of the excavation for their construction. This is especially difficult on tight urban sites, where the basement walls may be extended to the property lines.

2.4 UNDERGROUND BUILDINGS

Buildings are occasionally built entirely underground. However, the most common occurrence of underground buildings is with basements that extend horizontally beyond the perimeter of the building aboveground. In either case, the element added to the basement structure is that of a roof that forms a terrace or supports soil.

Placing a terrace or soil on a roof presents major structural and waterproofing problems. These are discussed extensively in Section 4.3. There is actually little comparison to ordinary roof construction, except for the waterproofing problem.

Truly isolated underground buildings have many additional problems regarding planning and construction. Entry and ventilation must be achieved, offering some special additional considerations for both the structure and general construction detailing.

Otherwise, the underground building has all the general problems encountered with basements.

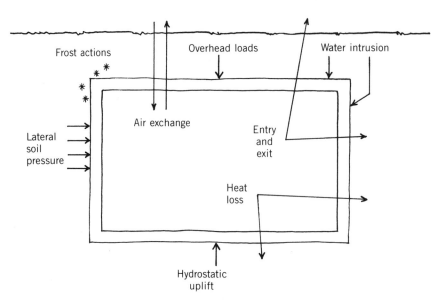

Figure 2.3   General concerns for below-grade spaces.

2.5 UTILITIES AND BUILDING SERVICES

Most buildings are served by connections to electric power, phone lines, and piped water and sewers. Many must also relate to gas piping, trash collection, and delivery services. Both building and site planning must relate in various ways to some orientation and connection to these services.

Water, sewer, and gas piping is generally underground and related to mains that are off the site, in a particular direction and at a specific elevation. Elevation is particularly critical for sewers, which must work by gravity flow. Frequently, in developed areas, there are separate sanitary sewers (building waste water) and storm sewers (roof and surface drainage runoff).

Placement of the building on the site and design of the building subgrade base and foundations must anticipate the facilitation of attachment to underground services. Where extensive regrading and/or site construction is planned, it must also be carefully coordinated with services, as well as existing property edges at streets and adjacent properties.

2.6 MISCELLANEOUS BELOW-GRADE CONSTRUCTION

In addition to the usual elements of foundations, basements, and underground services, various other items may be built into the ground on a site. Some such items are the following:

Vaults and manholes: For service to a site or in conjunction with general area development, providers of utilities may need to install some underground structures on the site. These take up underground space and also need access through surface-level covers.

Tunnels: For large projects, underground services may be placed in tunnels, permitting continuous access for alteration or service without disruption of the site surface. Disruption of drives, parking areas, or extensive terracing or landscaping can thus be largely avoided. Tunnels may also accommodate pedestrian or vehicular traffic, connecting underground building spaces to streets or subways, connecting separate buildings on the site, and so on. This is being increasingly done for building complexes in very cold climates.

Existing services: By previous arrangements, underground service mains, subways, tunneled drainage channels, or other elements may already be in place on an otherwise undeveloped site. This is most likely in highly developed areas, such as urban centers.

2.7 SITE STRUCTURES

Development of almost any building site involves some site construction, that is, some elements built outside the building. Sidewalks, drives, curbs, planters, and fences are most common. For sloping sites, earth berms, retaining structures, terraced slopes, and other elements for slope control and elevation changes are usually necessary. This may be modest for the average small project with the building occupying a major portion of the site. But it can also be quite extensive for large, parklike sites.

This book is basically about *building* construction, so the treatment of site construction is dealt with minimally, involving only the most common elements. However, the full management of the site as a resource and complete, well-coordinated design of the whole building and site development should incorporate all of the construction work.

Site construction of various forms is discussed in Chapter 4, and many typical elements are illustrated in the building cases in Chapter 7.

2.8 EASEMENTS

As previously mentioned for underground elements such as subways, prior arrangements may be in place to give certain rights of access or use to a property. One such arrangement is an easement, which reserves certain rights for someone else regarding a given property. Air rights for overhead power lines are one such item. Others include rights to extend roads, sewers, water mains, and so on. This does not always block the possibility of some site surface development, but it must be done with the risk of disruption in mind.

2.9 PLANTINGS

Development of plantings on sites may consist of preservation of existing materials, of insertion of new plantings to replace or augment the existing site, or of wholly new plantings. Whenever new materials are to be developed, there is usually some construction associated with the plantings. At the least, this consists of the preparation of the soils into which the plantings are placed for growth. In many instances this means the importing or moving of soils to provide the desired type of environment for the planting.

This book will not treat the general design of landscaping, but will consider some of the construction problems associated with plantings. Some major considerations in this regard are presented in the following discussions.

Adequate Soil

Some plants are hardy and can grow in many situations; others are quite sensitive and have stringent limitations if they are to flourish. The type of soil available on the site—or feasibly obtainable in the quantities required—must be considered in this regard. However, the permeability of sublevel soils, amount of local precipitation, and elevation of the groundwater level (free level of water in a dug hole) must be considered. Irrigation may augment the natural precipitation, but overabundant water is harder to correct for.

Providing for good soil conditions may become a major construction problem if both the topsoil and sublevel soils must be replaced, or if special subsurface drainage methods must be used. This can become quite extensive for large plants, such as trees.

Irrigation

Plants need water in some controlled amount to sustain growth. Needs may change over the seasons and the growth cycles of the plants and the normal conditions of seasonal precipitation to which some plant species have become accustomed. Cactus *expect* to be water-starved for some periods of time and will probably rot if continuously irrigated in show-draining soils. On the other hand, plants from coastal or tropical forest areas probably need some irrigation that does not let their roots ever fully dry out.

Development of an appropriate irrigation system is a whole design problem in itself which must relate first to the needs of the plantings it must sustain. However, it must also relate to the rest of the site and the site and building construction. Effects on basements and foundations must be considered. Runoff of excessive irrigation water must be achieved without impairment of site-related activities (such as walking into the building).

The general problems of water control on sites is discussed more fully in Section 5.5.

Root Growth and Plant Growth

Some plants grow to a limited size, and some are maintained at a shape and size by cutting back. If allowed to grow without limit, however, many plants will develop more or less continuously, both above and below the ground surface. Many buildings and sites have become overtaken by large shrubs and trees that were either badly chosen for their locations or simply not well maintained by frequent cutting back.

Often more difficult to control is root growth. Plants will send their roots in search of water (and air and nutrients), and the roots will go in any direction and grow to the necessary size to achieve their mission. Buried piping, pavements, and shallow footings for walls or planters may be overrun, lifted, or otherwise affected.

Protection of Plantings

In the forest and jungle, plants fight for survival, but in the captive site situation it is usually necessary to help most plantings by sheltering and nourishing them. Dividing of planted areas, isolating of sensitive plants that do not like crowded conditions, selective irrigation, and other methods may be required to keep the planted neighborhood happy.

Forming of the site surface and use of site constructions such as edging or planters—or even isolated, exterior containers—may be means for both protecting plants from each other and making individual care easier.

2.10 SOIL STRUCTURES

Undoubtedly, the major "element" in site construction is the site itself, as constituted by the soils that define its surface and immediate subsurface layers. Making of a site as a constructed object means working extensively with the site soils. This is a significant aspect of foundation design, pavement design, landscape design, and so on.

Besides providing support or encasement for various objects, soils are also used frequently for some forms of direct construction. Although topsoil, plantings, paving, or various ground covers may be used to develop surfaces, the general surface is usually developed by the underlying soils. Achieving this general site "construction" means working with soils as construction materials. It is necessary to have this view in order to appreciate the need to understand something about the structural character—and limitations—of soils.

Use of soils in various situations is discussed throughout this book. A general discussion of soils and their properties is provided in the Appendix.

3

Materials

Materials for building construction include traditional ones long in use and new ones emerging as recent technological developments. The following discussion treats primarily the most common and widely used materials for current building construction. Emphasis here is on basic construction; finish materials for surfacing abound in great variety, but will be treated only by general categories in this book.

3.1 WOOD AND WOOD FIBER PRODUCTS

The enduring large stands of forests in North America constitute a resource for wood for many purposes. For buildings, major uses are the following (see Figure 3.1):

Solid wood, sawn and shaped: This includes structural lumber plus elements used for trim, shingles, flooring, window and door frames, and finish wall surfacing.

Glued laminated elements: These include plywood, timbers made from multiple lumber pieces, and various other products.

Fiber products: These are produced with wood fiber as a major bulk ingredient and include paper, cardboard, compressed fiber panels, cemented shredded fiber units, and cast products with wood as a principal aggregate.

Wood is organic and is subject to decay over time (by simple aging or by decomposition), to rot, and to consumption by insects. A critical aging process occurs in the early weeks and months after a tree is cut down. For solid sawn pieces, a major property is the degree of retained moisture, which generally indicates the extent to which the wood has cured from its fresh-cut (green) condition. The concern is not for the moisture itself, but the the extent to which natural shrinkage of the material has occurred, as this is what mostly causes warping, splitting, and the general development of flaws, shape changes, and dimensional changes.

For site uses, wood is often left exposed for the attractiveness of its natural colors and grain patterns. Elements in the wood, such as knots—which are actually flaws—often give the exposed wood a richer texture. However, even "natural" wood is often protected by coatings to develop resistance to wear, enhance its color, resist staining, and so on.

While wood comes basically from renewable sources, the quickest renewal (fast growth) is achieved with trees that are functional primarily as fiber sources and not for solid sawn products. Large old trees, of the types used for the best solid wood products, are a dwindling source. For this and other reasons of economy, fiber

(a)

(b)

Figure 3.1   Wood elements in site construction (a) Heavy treated timbers used for
retaining walls. (b) Planters. (c) Fences.

(c)

Figure 3.1 (*continued*)

products and some forms of laminated products are steadily replacing solid wood in many applications. Similarly, fiber products are also displacing plywood for many uses involving low structural demand.

Wood is still the structural material of choice in the United States for most ordinary construction and is mostly displaced only when there is heightened concern for fire or need for some special property that is beyond the potential of the materials. Even if it eventually recedes significantly as a solid sawn product, wood will endure as paper and other fiber-based products.

Information Sources

National Forest Products Association

American Institute for Timber Construction

Western Woods Products Association

American Plywood Association

3.2 METALS

Metals are used extensively in building construction, although development of new synthetic materials works continuously to replace them in many common uses. As long as their basic properties (significantly: strength, stiffness, hardness, etc.) are critical to certain tasks (notably: hardware and major structural uses), and their cost is reasonable, usage generally continues.

Steel

Steel is the most widely used metal for construction, including both size and usage ranges that are greater than that of any other material (see Figure 3.2). Hardly any form of construction can be achieved without some steel products. Nails and other connectors for wood, reinforcement for concrete and masonry, and innumerable hardware items are required in every building.

The most impressive uses of steel, however, are for the large rolled beams and columns for high-rise building frames and the stranded cables used for tension structures. Even for structural applications, however, extensive uses are for modest elements, such as formed sheet steel decking.

When exposed to air and moisture, steel rusts—a progressive condition that eventually destroys the material completely. Large, thick pieces may take a long time to rust away significantly; therefore they endure for a reasonable time. The thinner the piece, or the more crucial it is to maintain its full cross section (for structural use, for example), the more serious it becomes to mount some rust prevention method. Rust prevention is usually achieved by coating the steel surface with other metals or a rust-inhibiting paint. Steel reinforcing bars may be coated with plastic where corrosion inside masonry or concrete is a concern. Special steels can also be produced (so-called *stainless* steel or various rust-arresting steels), but these are usually quite expensive and not all steel elements can be produced with them.

Steel structures are noncombustible and are used to replace wood construction where codes require a form of noncombustible construction. The light wood frame (2 × 4's et al.) can be duplicated in light-gage steel (formed sheet) elements for wall framing, for example. However, steel heats rapidly and steel elements soften quickly, so some protection is required for steel structures where any significant fire rating is required.

Information Sources

American Institute of Steel Construction

American Iron and Steel Institute

Aluminum

Aluminum is used only sparingly for structures, as its cost is prohibitive. However, it has considerable corrosion resistance in some applications, weighs about one-third as much as steel, and can be extruded or cast into complex, sharp-cornered forms— giving it an edge for usage in various situations.

Aluminum is used in buildings most extensively for window and door frames and for decorative trim or framing for modular ceiling systems, light fixtures, and various finish, furnishing, or service elements. Aluminum foil is used as a moisture barrier and/or reflective material in various wall and roof assemblies to enhance the barrier functions of the building enclosure.

Information Sources

The Aluminum Association

Architectural Aluminum Manufacturer's Association

(a)

(b)

Figure 3.2 Steel elements in site construction: (a) Fences. (b) Site lighting.
(c) Railings.

(c)

Figure 3.2 (*continued*)

Other Metals

Other metals are used for special purposes, where their properties are significant. Flashing was done extensively in the past with thin sheets of copper, lead, tin, and special alloys, but it is now also extensively achieved with various plastic laminates. Decorative elements of chrome, bronze, copper, or other metals are used, but are often achieved with thin coatings over a base metal or with thin films laminated in a plastic structure.

Reflective glazing is sometimes achieved with ultrathin silver or gold films adhered to glass or, now, more often laminated into a sandwich of glass and plastic.

Many metals are still used where their functional properties are significant and no existing replacement is fully competitive. However, continued developments in materials technology steadily bring forth new materials, such as fiber-reinforced concrete and high-strength plastics, that bump metals out of general usage for building construction.

3.3 CONCRETE

Concrete is a material consisting of loose, inert particles (called aggregate) that are bound together by some infilling binder. That definition could be extended to include many mixtures, but the principal ones significant to building construction are the following: (See Figure 3.3.)

Structural concrete: This is the material produced for sidewalks and streets as well as foundations and building frames. Aggregate is typically locally available gravel, crushed stone, sand, clam shells, and so on, selected in a range of sizes so the little pieces fill the spaces between the big ones. The objective is to have the aggregate take up as much of the volume as possible, producing a tightly packed, dense mass with very little room for the binder, which consists of portland cement and water. This basic mix can be modified in various ways by additions, called admixtures, to change the color, weather resistance, water penetration resistance, and so on.

Synthetic, lighter-weight aggregates may replace the gravel to reduce the unit weight by as much as 40% or so and still produce structural resistance of significant amounts.

Insulating concrete: This is ultralightweight concrete, produced with portland cement and some very lightweight aggregate—such as wood fiber or some low-density mineral substance, such as vermiculite or perlite. The mix is lightly foamed with as much as 20% air, and the unit density can be as low as 30 pcf, compared with 110–150 pcf for structural concrete. This material may be used for fire protection of steel elements, but it is mostly used as fill on roof decks.

Fiber concrete: Research and experimentation have resulted in the use of various forms of concrete with fiber materials in the mix. Their general purpose is to add some additional tensile capacity to the concrete, reducing or eliminating the need for the usual steel reinforcement used to alter the tension-weak character of the concrete. Roof tiles, siding, and pavements are some of the current applications for this form of concrete.

Gypsum concrete: Gypsum can be substituted for portland cement in plaster or concrete. A lightweight roof deck material is produced with gypsum and an aggregate of wood chips. This is an insulative material with significant strength and stiffness for roof structural applications in controlled situations. Priority systems consist of combinations of gypsum concrete cast over modular metal supports and forming units, the forming units developing finished ceilings in some applications.

Bituminous concrete: Substitute a hot bitumen (coal tar, etc.) for the portland cement and water and you have asphalt pavement. Not much used for buildings, but a significant material for site development.

In most applications, structural or not, some compensation must be made to overcome the tension-weak character of the concrete. Steel reinforcement, in the form of rods or welded wire mesh, is the usual solution. The concrete encasement needs to protect the steel to prevent rusting or insulate it from fire. Other means for improving tension resistance include the use of fiber materials (as mentioned previously) and prestressing, which consists of inducing compression in structural concrete members before the occurrence of loading that produces tension. Prestressing is usually induced by high-strength steel strands stretched inside the concrete and anchored or bonded to the concrete. Maintaining the stretch of the steel develops compression in the concrete.

Concrete must be cast in forms and cured over some period of time after it is mixed and placed in the forms. Hardening results from a chemical action between the water and cement, and it takes some time to occur in ordinary situations. The cast concrete must be kept moist and at a reasonable temperature during the curing period.

Mixing, transporting, casting, finishing, curing, and reinforcing or prestressing of concrete represent major cost factors. Cost reduction of these items is often more significant than attempts to make the concrete mix itself cheaper. Better to go for the best mix and find ways to reduce other costs—particularly the time-consuming and labor-intensive ones.

Concrete is generally the most durable material for ground-contacting construction, and is thus extensively used for foundations, basement walls and floors, pavements, and other site construction. It is also generally weather resistive and can function reasonably in an exposed condition as a raw, unprotected material. Finally, it

(a)

(b)

Figure 3.3 Concrete elements in site construction: (a) Sidewalks, curbs, and steps. (b) Ramps and bridges. (c) Freestanding wall. (d) Site furniture. (e) Precast cribbing for a steep slope.

(c)

(d)   **Figure 3.3**   (*continued*)

(e)   **Figure 3.3**   *(continued)*

is considerably fire resistive. All of this frees it from the need for the many concerns for protection often required for wood and steel.

Because of its functional capability of being exposed—on both the exterior and interior—concrete surfaces are frequently in view, mostly for walls and the underside of spanning roof and floor structures. Where this is the case, and appearance is of some concern, it is necessary for the designer to deal with the various factors that influence the development of surface form and texture. These factors are discussed at length in the following chapters.

Information Sources

American Concrete Institute

Portland Cement Association

Prestressed Concrete Institute

3.4 MASONRY

The term *masonry* covers a range of construction that overlaps the areas of tiling, cladding, and precast concrete. Traditional forms of brick and stone masonry can still be obtained, but other processes and materials have largely displaced them. Most structural masonry for bearing walls, shear walls, or foundation walls is now produced with units of precast concrete (concrete blocks—called CMU, for concrete masonry unit). What appears to be a brick wall these days is often a frame or CMU structure with thin "brick" tiles adhesively bonded to a backup material and mortar joints simulated with a filler material.

Stone is mostly used decoratively these days. Fieldstone is developed as either veneer or facing on other structures. Cut stone (granite, marble, etc.) is cut in increasingly thin slices to form facings over structures of other materials.

In the face of these trends, the crafts of stonemasonry and bricklaying slowly decline and become ever less obtainable. The enduring popularity of the form of construction (or at least its appearance) keeps it alive—but increasingly as illusion or metaphor, rather than as fact.

Classic masonry consists of some inert units bonded into a contiguous mass with a binder between the units (see Figure 3.4). The binder is usually mortar—similar in form to that used in ancient times, but quite different today in its ingredients and properties. Units may be stone—cut or unprocessed—bricks of formed and fired clay, precast concrete (mostly as hollow blocks), or other elements, such as hollow tiles of gypsum or fired clay.

It is desirable to have masonry consist mostly of the units, with a minimum of mortar in thin joints between the units. It is best to rely on the mortar only for compressive resistance, although it is desirable to have it adhere well to the units to bond them together for general stability of the construction.

Overall, the strength and stability of masonry construction depends on the units, their arrangement, the mortar joints, and—above all—the craft of the workers who prepare the units, mix the mortar, lay up the units, and protect the construction while the mortar hardens. When it all works, it is great, but in modern times the biggest drawback to the use of classic masonry construction is its dependency on the craft of the workers.

Masonry units must be closely fit together, so the size and shape of the units, the patterns of their arrangement, the general form of the construction they are forming, and the required edges, corners, openings, intersections, supports, and other form-related concerns must be carefully developed. This is especially critical for work exposed to view. It is also most critical for CMU construction; bricks and stones can be cut to trimmed sizes and shapes, but concrete blocks must be used in whole-piece form. Special CMU shapes and sizes are provided, but some dimensional and modular control is necessary.

Like concrete, masonry is tension-weak, which needs consideration in structural applications. The strongest masonry structures are usually those developed as *reinforced masonry*, which uses steel rods in a manner similar to that for reinforced concrete. This is of greatly heightened concern for structures required to resist major earthquakes or violent windstorms.

Information Sources

Brick Institute of America

Clay Products Association

National Concrete Masonry Association

(a)

(b)

Figure 3.4 Masonry elements in site construction: (a) Concrete block foundation walls. (b) Stone and concrete retaining wall.

3.5 PLASTICS

Plastics are used extensively in buildings as solid formed elements for trim and hardware, as film for moisture barriers or in laminates, as foamed insulation, and for many forms of coatings. As in its early days, plastic is often used to imitate other materials. (Is it real or is it *plastic*?) Thus wood paneling is often a photograph of wood embedded or laminated into plastic and mounted on a compressed wood fiber panel; wood strip siding is duplicated in form and appearance with formed vinyl sheets; metal trim or surfacing is actually thin metal film over plastic; brick, Spanish clay tile, mosaic tile, marble, and just about anything can be imitated with plastic in Disneyland style.

Basic chemistry, forming methods, and combinations of plastic with other materials can be varied extensively to respond to many demands. Potential problems include concerns for fire resistance, low strength or stiffness, wear resistance, stability over time, and possible toxic effects.

Many paints, varnishes, and other coatings have a plastic base. Multiple coatings may produce combined effects, with each coating adding something additional to the total effect.

Information Source

Society of the Plastics Industry

3.6 SOILS

Undoubtedly, the major material used for site development is soil. Building in and on the ground literally means using soils as construction materials. The general nature of soils—including their structural properties—must be reasonably understood by any designer involved in site, foundation, or below-grade construction.

For landscaping development it is also necessary to understand the nature of soils related to establishment and nurturing of plantings. Where extensive planting is to be developed, an early issue to be settled is the potential for using existing site materials and the means necessary to preserve them for final landscaping work.

This book deals essentially with construction, so we will not attempt to fully treat the subject of planting design, except for some aspects related to construction problems. The general engineering concerns for soils, however, are treated in the discussions in the Appendix.

Information Source

Simplified Design of Building Foundations, 2nd ed., James Ambrose, New York: Wiley, 1988.

3.7 WATERPROOFING MATERIALS

Water control is a general problem in building construction, and below-grade construction has very special concerns in this regard. Various materials are used in the form of membranes, coatings, sealants, flashing, and miscellaneous joint-sealing devices. The range of materials is considerable, including metals, plastics, and various bituminous compounds.

This subject is best addressed in terms of various specific types of problems and particular construction elements. It is treated in various sections in Chapters 4 and 5 and in several of the building case studies in Chapter 7.

Information Source

Time-Saver Standards for Landscape Architecture, Charles Harris and Nicholas Dines, New York: McGraw-Hill, 1988.

3.8 PAVING MATERIALS

Surfacing of the ground can be achieved with raw soil, with general plantings (grass, etc.), or with a variety of pavings. (See Figure 3.5.) The desire or need for paving usually arises in anticipation of some form of traffic, and a primary concern must be for the support of the type of traffic. Fine gravel or pulverized bark may surface a lightly trodden garden pathway, but something serious is required for heavy trucks on a daily basis. Pavings for various purposes and with various materials are discussed in Section 4.7 and illustrated in Chapter 7.

(a)

Figure 3.5   Paving materials: (a) Concrete, brick and asphalt. (b) Brick, loose-laid on sand. (c) Open concrete units for drainage control.

(b)

(c)

Figure 3.5 *(continued)*

3.9 MISCELLANEOUS MATERIALS

Many materials can be used for various purposes for building construction. Some are used in unique applications; some are used as substitute for more traditional materials—for cost savings, or for some enhanced property that is superior to the material it is replacing.

Paper and Other Fiber Products

Paper is used with various elements of building construction. Two common uses are the facings for gypsum drywall units (a sandwich with a gypsum plaster core) and the backup material for stucco. Paper can be coated—with plastic, but also with aluminum foil, wax, or bitumen. It can also be laminated; for example, with glass fiber strands for reinforcement.

Cardboard in solid form is more or less thick paper, intermediate between ordinary paper and various forms of compressed wood fiber panels. Many variations, including ones with coatings or lamination, are possible. Concrete forming, such as that for round columns, is sometimes accomplished with heavy cardboard elements.

Corrugated cardboard is a sandwich of alternating layers of flat and pleated or corrugated paper, resulting in a material with considerable strength and stiffness but relatively light weight. As with other paper products, special coatings and reinforcements are possible.

Composite Materials

Composite elements are those in which two or more different materials are blended in a way such that the materials retain their individual identity but share some effort with the other material(s) with which they are blended. Examples are reinforced concrete (concrete plus steel) and laminated glazing (glass plus plastic). Increasing use is being made of composites in building construction. In some cases these produce entirely new elements with many new potential possibilities. In other cases they simply permit an extended usage of old, familiar materials beyond their previous limitations.

4

Elements

General concerns for basic elements of construction are discussed in Chapter 2. This chapter presents materials relating to specific use of products and systems for site and below grade construction. Some special concerns for various properties of the construction are discussed in Chapter 5. Systems for development of whole buildings and sites are discussed in general in Chapter 6 and illustrated for particular cases in the examples in Chapter 7.

4.1 BASEMENTS

Basements were quite common in the past, a principal purpose being the housing of equipment for either hot-air, hot-water, or steam-heating systems, all of which needed gravity flow effects for functioning. Heat energy sources consisting of massive quantities of wood and/or coal had to be stored and be near to furnaces. Also, the generally unheated basement was useful for storage of perishable foods before refrigeration. And then there was the idea, if not the reality, of a secure place to go in windstorms.

Emergence of forced air, counterflow heating and air conditioning, energy sources of electricity, gas and oil, the common refrigerator and food freezer, and other factors have made the basement less necessary. Still, there are enduring factors and some new ones that continue the use of basements.

In cold climates earth sheltering is an energy-saving device, and foundations must generally be extended some depth into the ground because of frost problems. In very tight site situations, especially dense urban locations, use of basement space may be an efficiency of planning. A building may be linked to others in a complex or to facilities for entry, waste removal, parking, ground transportation, or emergency shelter by underground connection through basements and tunnels. Pedestrian circulation below ground may avoid street traffic or exposure to outdoor temperature extremes.

In any event, basements are here to stay, and their various problems must be dealt with in the full design development of a building and its site. The following are some fundamental considerations for development of basement construction.

Structure

Basements essentially consist of side walls and a floor. The top of a basement is usually the bottom (ground floor) of a building—having basic identity and a primary purpose to achieve that function. However, basements may extend beyond the site footprint of the aboveground building and become effectively simply underground buildings. This adds the concern for a grade level or below-grade roof—a special problem considered in Section 4.3. Here we consider primarily the wall and floor structure.

Most basements are achieved with concrete construction, typically as sitecast structures with formed walls and slabs on prepared soil bases. The typical wall construction is that shown in Figure 4.1(a). Required wall thickness and reinforcement derives from the basement depth and wall-spanning dimension, with the wall typically functioning as a vertical-spanning slab in resisting the lateral soil pressure.

As in other situations, concrete walls and slabs are reinforced with two-way sets of reinforcing bars to resist temperature expansion and concrete shrinkage effects, as well as structural effects. If upward hydrostatic pressures from groundwater exist, floor slabs may have major structural functioning as spanning elements; otherwise, they basically just lie on the ground beneath them.

Basement walls also frequently serve functions as general elements of the building base and foundation. Exterior building bearing walls will transfer their loads finally to basement walls, making them serve as bearing walls or grade walls (spanning between isolated foundations). These problems are discussed further in Sections 4.3 and 4.4.

Where masonry construction is generally favored, small buildings may have masonry walls—a common form being that with CMUs (concrete blocks), as shown in Figure 4.1(b). Although somewhat weaker than concrete walls, these may be quite adequate if built properly. Presently, construction is usually achieved with the CMU cavities filled with concrete and some use of steel reinforcement. This form of construction is extensively used for residential construction in areas where basements are most common—the eastern and northern parts of the United States.

In most regards, basement floor slabs of concrete are essentially similar to other paving slabs, although some special problems may occur. Basement floors in general are discussed more fully in the next section.

Basements, whether totally under a building or simply underground, must respond to various concerns for the complete development of the construction. The general relationships were described in Section 2.3 and illustrated in Figure 2.3. In addition to the raw concrete or masonry structure, there are usually several other features, as described in the following discussions.

Water Control

A major general problem for basements and underground spaces is that of water. This includes possible concerns for the following:

Groundwater: All soil contains some moisture, or has the potential to. Moisture may be ever-present or occur mostly only during periods of extensive precipitation. Intrusion of moisture is a common problem for underground spaces. Basement walls and floors must resist this general moisture intrusion. If the free water level in the soil is above the level of the basement floor, the problem is considerably aggravated.

Precipitation runoff: A general site design rule is to have the ground surface around a building slope away from the building, hopefully reducing any potential problems for soaking the basement walls from rain or melting snow. However, roof

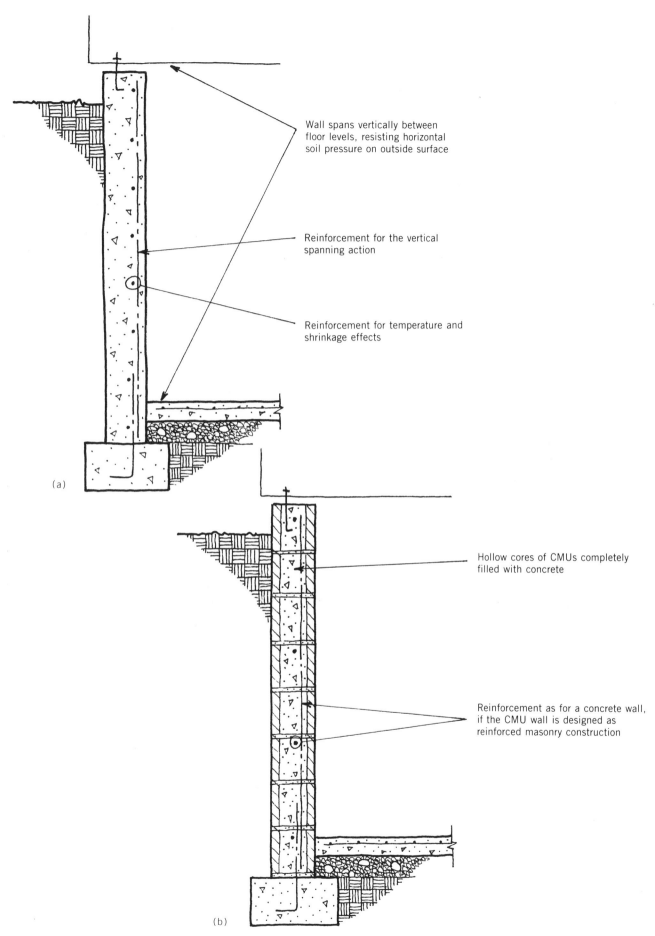

Wall spans vertically between floor levels, resisting horizontal soil pressure on outside surface

Reinforcement for the vertical spanning action

Reinforcement for temperature and shrinkage effects

(a)

Hollow cores of CMUs completely filled with concrete

Reinforcement as for a concrete wall, if the CMU wall is designed as reinforced masonry construction

(b)

Figure 4.1 Basement wall structures: (a) Sitecast concrete. (b) Concrete block.

33

drainage may aggravate this. And, even if roof drainage is controlled, water runs down the face of exterior walls, cascading at the bottom at the building edge.

Irrigation: Building edges are popular locations for plantings, which typically need continuous irrigation. This inevitably produces the situation that all efforts regarding precipitation try to avoid.

Building water use: Piped-in building water typically enters a building through its basement. Hot water is typically produced in the basement. Hot water or steam boilers and circulating equipment are usually in basements. Building wastewater and sewerage, including any interior roof drainage, usually end up in the basement before leaving the building. This all presents various details for the construction— including many penetrations of the basement walls and floor. And, if Murphy's law prevails, leaks of various piping or equipment *will* occur.

Figure 4.2 shows some of the enhancements of the raw structure shown in Figure 4.1, many of which relate to water problems. The seriousness of these problems and the extent of provisions for them may well relate mostly to the use of the building and particularly to the use of the basement spaces.

General Enhancement of Occupied Spaces

Basements may contain essentially only equipment and storage spaces. Humans may enter—but do not *occupy*— the space, in the building code's meaning of the term. If basement spaces are truly occupied, however, there will typically be heightened concern for water, thermal conditions, ventilation, and general finishing of interior surfaces.

Basement walls can be insulated in a variety of ways, depending on the primary effects to be obtained. General energy conservation is one concern; thermal comfort of occupants is another. Insulation on the outside generally retards the effects of ground frost and cold air conditions at the ground surface. Insulation on the inside further reduces the cold, radiant effects of wall and floor surfaces.

Finishes of all the usual types can be achieved for floors and walls, if proper attention is given to installation details and choices of materials. Attachment to concrete or masonry surfaces is a first item to be dealt with. Enduring potential water problems comprise another.

Second to roofs, water problems in basements are probably one of the nastiest problems in building construction. This area deserves a lot of attention for successful development of building construction.

4.2 BASEMENT FLOORS

Most basements are one story, and their floors are concrete slabs on a prepared soil base. Basements for large buildings may be multistory, in which case other floors are essentially not different from framed structures in aboveground buildings. This section deals essentially with the paved slab floor.

Structure

Pavements in general are discussed in Section 4.7, which includes a general treatment of reinforced concrete slabs on grade. All concrete paving slabs are essentially

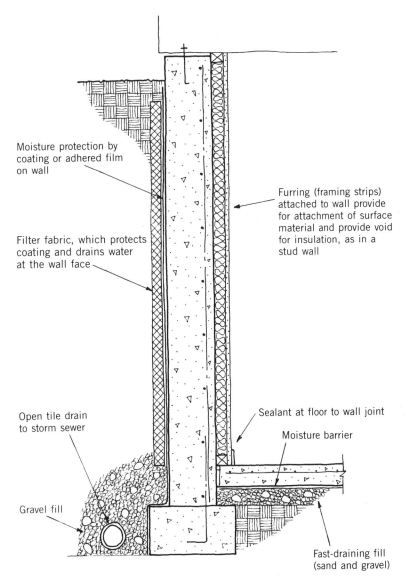

Moisture protection by
coating or adhered film
on wall

Furring (framing strips)
attached to wall provide
for attachment of surface
material and provide void
for insulation, as in a
stud wall

Filter fabric, which protects
coating and drains water
at the wall face

Open tile drain
to storm sewer

Sealant at floor to wall joint

Moisture barrier

Gravel fill

Fast-draining fill
(sand and gravel)

Figure 4.2   Basement wall construction.

similar in a structural sense, although use as building floors, basement floors, walks, driveways, or other specific applications bring particular concerns. The following discussions deal mostly with the various nonstructural concerns.

A special situation for basement floors, however, is the possibility for upward pressures, due to either expansive-type soil or water pressure. At an extreme, this may require design for a major loading, requiring a special spanning structure instead of the usual semidormant slab lying on soil. A form of such a structure is shown in the basement for building 7 in Chapter 7.

Water Control

The minimum effort for water control usually consists of providing for some resistance to simple upward migration of moisture through the normally quite absorbent concrete slab. Pouring the concrete on top of a granular (sand and gravel) base—as is done mostly for all pavements—helps reduce simple upward percolation effects.

Placing a nonpermeable barrier (plastic film, treated construction paper) between the concrete and base further restricts the upward moisture movement.

If water is likely to accumulate under the floor due to periodic high groundwater conditions, it may be possible to use an underfloor drain system in conjunction with a perimeter footing drain system and to feed this to a sump which is used to evacuate the accumulated water. Figure 4.3 shows the general form of such a system.

If there is an enduring hydrostatic head of water in the ground above the level of the basement floor, a much more serious waterproofing must be achieved—essentially the same as for a flat roof, with the added problem that the water is under pressure. Buildings with deep basements near large bodies of water are such a situation. This is a major singular construction problem, and we will not attempt to treat it here. The much more common situation is the simple need for a modest reduction of moisture intrusion.

As mentioned previously, moisture is generally treated with more concern when basement spaces are occupied by people. This includes concerns for environmental conditions for the occupants and the protection of the additional floor and wall finish materials normally used in this situation.

Floor Finishes

Unoccupied basements usually have the raw, unadorned concrete slab surface as a floor finish. The surfaces may be finish troweled to a hard, smooth finish and possibly treated to reduce slippery conditions when wet, but additional, separate finish materials are not common. An exception may be the provision of a nonconductive finish in areas with extensive electrical equipment.

Finishes with additional materials are usually reserved for occupied spaces. Virtually all options available generally for finishes applied to concrete substructures are possible. An additional concern, which may affect construction details but not necessarily the choice of actual finish materials, is the desire to reduce heat loss (or cold gain) from the normally cool ground (from about 50° to 60° in most regions). A carpet with a thick pad may help this, but a thin tile applied directly to the concrete slab will not.

4.3 BELOW-GRADE ROOFS

When a space exists truly underground, all of the problems described in preceding sections for basement walls and floors are present. A major additional concern, however, is that for an overhead structure that is *not* simply a floor of a building above. This structure, which we call the roof of the underground space, may support soil of some thickness or be paved as a terrace or plaza surface. With an extensive underground construction and major landscape development, all possible overhead conditions may exist.

Structure

Compared to ordinary roofs, those for underground spaces usually carry many times the total gravity loads. The construction of the structure itself is likely to be among the heaviest we use—sitecast concrete. No light trusses, space frames, or fabric structures here! Add to that either a heavy load of soil or the high live load usually required for a terraced roof area or a plaza (150–250 psf in most codes).

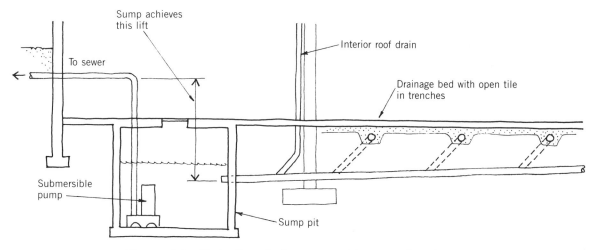

Figure 4.3 Underfloor drain system with sump pit and pump.

A further concern is often that of restricting the total depth of the structure in order to keep the depth of the underground construction to a minimum. All of this conspires to generally keep spans short and try for some efficiency in the spanning systems if possible. The usual options for sitecast concrete floor systems are all possible, as shown in Figure 4.4. Among these, the two-way flat slab and two-way waffle systems usually offer the smoothest underside and least overall depth and are favored, especially if the underside of the structure is left exposed.

Water Control

A roof is a roof, and the generally basically flat roof of an underground structure is not significantly different in many regards from any flat roof. A totally watertight membrane is required, together with all flashing and careful sealing of any penetrations. The membrane must be protected, especially here, and some insulation and a vapor barrier may be indicated.

Drainage is somewhat different here than for a conventional roof, but not any less indicated. The form of drainage will relate somewhat to what is above the roof—soil or paving—and probably to the development of a total drainage system for the whole underground construction. As in other situations, the roof must likely be drained essentially by gravity flow, involving considerations for the total slopes required and the vertical dimensions of the construction necessary to achieve drainage and removal of water.

As for walls and floors, concern for the watertight security of the underground space may be heightened if the space is occupied by people. However, there is hardly any middle ground for a watertight roof—it is or it isn't, so this is less a variable issue than it is for basement walls and floors in ordinary circumstances.

Earth-Covered Roofs

An especially critical water control problem is that of the roof supporting earth—and usually plantings. The earth may get wet from precipitation alone, but the plantings

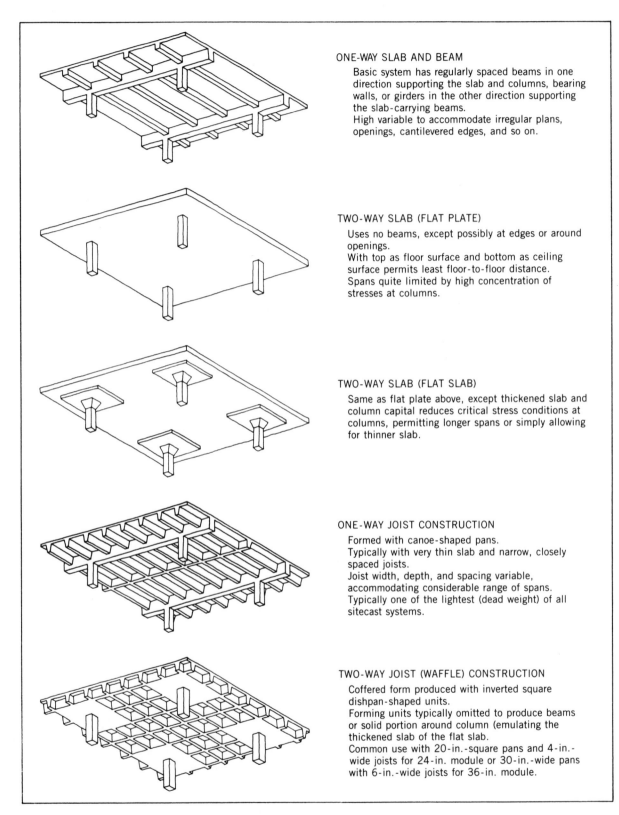

ONE-WAY SLAB AND BEAM

Basic system has regularly spaced beams in one direction supporting the slab and columns, bearing walls, or girders in the other direction supporting the slab-carrying beams.

High variable to accommodate irregular plans, openings, cantilevered edges, and so on.

TWO-WAY SLAB (FLAT PLATE)

Uses no beams, except possibly at edges or around openings.

With top as floor surface and bottom as ceiling surface permits least floor-to-floor distance.

Spans quite limited by high concentration of stresses at columns.

TWO-WAY SLAB (FLAT SLAB)

Same as flat plate above, except thickened slab and column capital reduces critical stress conditions at columns, permitting longer spans or simply allowing for thinner slab.

ONE-WAY JOIST CONSTRUCTION

Formed with canoe-shaped pans.

Typically with very thin slab and narrow, closely spaced joists.

Joist width, depth, and spacing variable, accommodating considerable range of spans.

Typically one of the lightest (dead weight) of all sitecast systems.

TWO-WAY JOIST (WAFFLE) CONSTRUCTION

Coffered form produced with inverted square dishpan-shaped units.

Forming units typically omitted to produce beams or solid portion around column (emulating the thickened slab of the flat slab.

Common use with 20-in.-square pans and 4-in.-wide joists for 24-in. module or 30-in.-wide pans with 6-in.-wide joists for 36-in. module.

Figure 4.4   Common forms of sitecast concrete spanning systems.

will require continuous moisture, so the water condition will be ever-present. Drainage must be effective to prevent a saturated, rot-inducing condition for the plant roots.

Another major consideration in this situation is the provision of sufficient depth of earth fill for sustaining of the planting. This may be minimal for grass, but of major proportions for trees. To save depth, large plants and trees may be placed in special planters integrated into the space of an otherwise not so deep underground construction.

Figure 4.5 shows some of the features of an earth-covered underground structure, sustaining only minimal plantings in this case.

Paved Roofs

When close to the surface, underground spaces may have paved roofs—forming terraces or plazas for buildings, or simply parts of a general site development. Extensive underground parking garages are developed in many locations with parks, squares, or other open spaces on their roofs. Paving here is not essentially different from other outdoor paving, except for the developed base as shown in Figure 4.6. The sub-base and general support must be generally integrated with the overall construction of the roof of the space below.

The same underground space may have both earth for plantings and paving on different areas of the roof. This calls for some coordination of the overall dimension for development of the two types of surfaces, if the roof is constructed essentially flat.

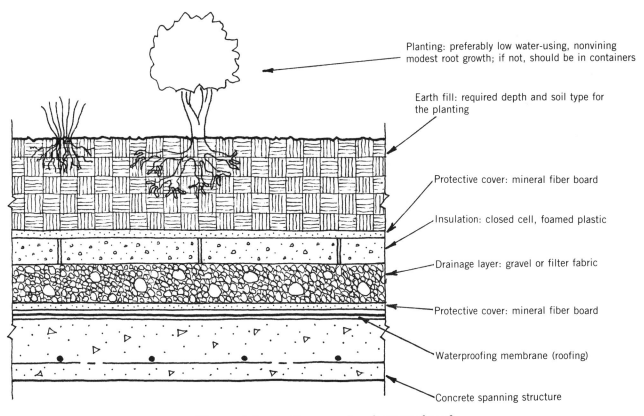

Planting: preferably low water-using, nonvining modest root growth; if not, should be in containers

Earth fill: required depth and soil type for the planting

Protective cover: mineral fiber board

Insulation: closed cell, foamed plastic

Drainage layer: gravel or filter fabric

Protective cover: mineral fiber board

Waterproofing membrane (roofing)

Concrete spanning structure

Figure 4.5 Earth-covered underground roof.

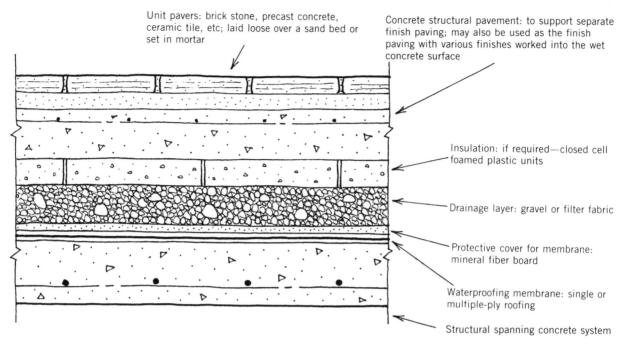

Unit pavers: brick stone, precast concrete, ceramic tile, etc; laid loose over a sand bed or set in mortar

Concrete structural pavement: to support separate finish paving; may also be used as the finish paving with various finishes worked into the wet concrete surface

Insulation: if required—closed cell foamed plastic units

Drainage layer: gravel or filter fabric

Protective cover for membrane: mineral fiber board

Waterproofing membrane: single or multiple-ply roofing

Structural spanning concrete system

Figure 4.6   Underground roof with paving.

Information Sources

Building Underground, Herb Wade, Emmaus, Pa.: Rodale Press, 1983.

Earth Sheltered Homes: Plans and Designs, Underground Space Center, University of Minnesota, New York: Van Nostrand Reinhold, 1981.

4.4 SHALLOW FOUNDATIONS

Shallow foundation is the term usually used to describe the type of foundation that transfers vertical loads by direct bearing on soil close to the bottom of the supported structure and a short distance below the ground surface. The three basic forms of shallow foundations are the continuous strip wall footing, the isolated footing for a single column or post, and the raft or mat foundation that consists of one big pad used for an entire large structure (such as a whole building).

The behavior of bearing footings relating to soil properties is discussed in the Appendix. Figure 4.7 illustrates a variety of elements ordinarily used in bearing foundation systems.

Wall Footings

Wall footings include footings for building walls, retaining walls, and tall fences built as continuous strips (masonry mostly). In addition to performing soil stress resolution functions, a wall footing generally serves as a construction platform for the supported wall. It must thus be related in form and dimensions to the wall and possibly provide for anchor bolts, reinforcement dowels, or other items necessary to the installation of the wall. Where total vertical loads are low in magnitude and soil is adequate, these requirements may effectively dictate the general form and dimensions of the footing.

For very large structural walls, on the other hand, footings may be quite wide and be designed for major cantilever bending forces due to their projection beyond the

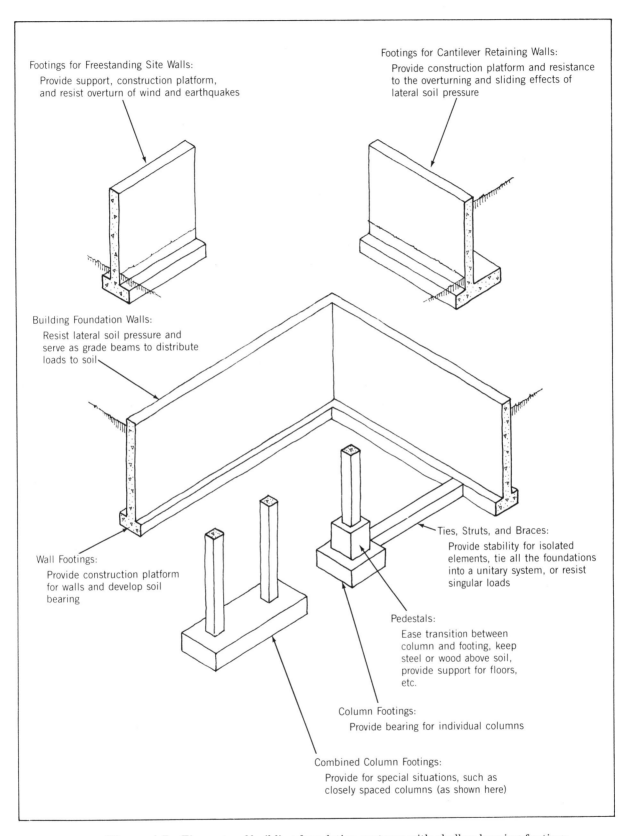

Footings for Freestanding Site Walls:
Provide support, construction platform, and resist overturn of wind and earthquakes

Footings for Cantilever Retaining Walls:
Provide construction platform and resistance to the overturning and sliding effects of lateral soil pressure

Building Foundation Walls:
Resist lateral soil pressure and serve as grade beams to distribute loads to soil

Wall Footings:
Provide construction platform for walls and develop soil bearing

Ties, Struts, and Braces:
Provide stability for isolated elements, tie all the foundations into a unitary system, or resist singular loads

Pedestals:
Ease transition between column and footing, keep steel or wood above soil, provide support for floors, etc.

Column Footings:
Provide bearing for individual columns

Combined Column Footings:
Provide for special situations, such as closely spaced columns (as shown here)

Figure 4.7 Elements of building foundation systems with shallow bearing footings.

wall faces. In addition to simple response to gravity loads, footings may need to resolve major horizontal sliding or overturning effects, as occur with tall retaining walls or freestanding walls.

In some situations, for walls that lack such capabilities, footings may need to function as spanning beams due to uneven ground conditions along their lengths. For all the necessary structural functions, footings must be adequately sized and reinforced.

Column Footings

While wall footings may be unreinforced laterally when they are barely wider than the supported walls, column footings must typically be reinforced in two directions for the cantilever effects from the concentrated column loading at their centers. For tall buildings or long-span structures, these usually square footing pads may be quite large in plan, thick, and heavily reinforced.

Pedestals

Pedestals, consisting of short columns or piers, may be used with column footings for various purposes. A frequent need is that of keeping the supported column above ground, as is the case for wood and steel columns. Column footings must typically be located some distance below the ground surface to assure good soil bearing, so this is a common problem.

Pedestals may also be used as transitional elements to help the transfer of force from the column to the soil. Heavily loaded columns typically have quite concentrated strengths. This may require a very thick footing to avoid a punch-through effect, and a pedestal may ease this transfer. Stepped pyramids of stone or masonry were constructed in the past to achieve this.

Pedestals may be constructed of cast concrete or masonry, depending on the load magnitudes and other general construction. They are basically just short columns, so conventional reinforced concrete or masonry column design is possible. However, when they are not more than three times as high as they are wide, they can use simpler forms of construction and reinforcement.

Combined Footings

In various situations, columns may be supported in groups by a single footing. Some reasons for this are the following.

Closely Spaced Columns

If required footings are large in plan dimensions, it may be difficult to use separate footings for closely spaced columns. One solution may be to use oblong plan-shape footings, rather than the usual square. This is usually reserved for situations where columns are close to some other construction, such as an elevator pit. Two columns close together are more typically supported by a single footing, which is designed like a beam between them.

Cantilever Footings

Cantilever or strap footings are used where an exterior building column occurs close to the property line and the usual column footing at the building edge would project off the property. This may also be the case where a new building is built virtually up against an adjacent one, even though on the same property.

Sensitive Supported Structure

Occasionally a structure or some supported equipment has great sensitivity to movements, and differences in settlement of separate footings may be critical; in such cases, a single footing may be effective.

Pole Foundations

A special foundation is that in which a timber pole is placed in an excavated hole. This is most often used for signs, overhead transmission lines, and fence posts, but also sometimes for aboveground decks or even buildings. The pole bottom essentially bears directly on the soil, although a compacted gravel or poured concrete pad may be placed at the bottom of the hole.

Figure 4.8 shows some possible forms for pole foundations. Details of the supported structure and general development of the site will affect choices for treatment at the ground surface around the pole. Effects of lateral forces may also influence some concerns for the depth of the hole, the general anchorage of the pole in the hole, and the treatment at the ground surface. A confining concrete pavement around the pole will drastically change its lateral stability.

Information Source

Simplified Design of Building Foundations, 2nd ed., James Ambrose, New York: Wiley, 1988.

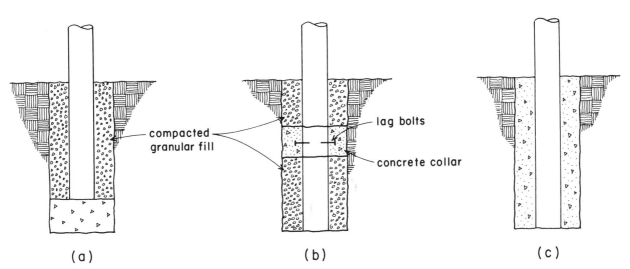

Figure 4.8   Methods for achieving pole foundations: (a) With a concrete footing and compacted gravel fill; in very soft soil. (b) With a concrete collar for more positive lateral bracing and anchorage. (c) With concrete-filled excavation for maximum bracing.

4.5 DEEP FOUNDATIONS

Far and away the cheapest, simplest, most common foundation is the shallow bearing footing. However, the use of bearing pressure on yielding soil has some limitations that must be recognized; these include the following:

Soil strength: Load capacity of soils, even the strongest ones, is limited. Good structural wood, for example, typically has a compression strength of around 2000 lb/sq in.—weak in comparison to steel or superstrength concrete. But a soil is considered superstrength if it can sustain as much as 10,000 lb/sq ft, or about 70 lb/sq in. Excessive loading, from heavy construction, tall buildings, or long-span structures, can require footings with massive plan areas to keep soil pressure low.

Settlement: Soil is compressible; the greater the pressure, the more the deformation, which accumulates as downward vertical movement (settlement). With large deposits of clays, settlements may be progressive over time, accumulating in *feet,* not inches.

Instability: Soils, with discrete particles not generally tightly bonded or cemented, are subject to erosion, consolidation, organic or general chemical decomposition, and massive shrinkage or expansion due to major changes in water content.

For these or other reasons, use of deep foundations may be favored. Actually, the simple reason for choosing what is almost always a more expensive foundation system is usually that the bearing capacity needed is not available at the location where bearing footings would need to be located.

Many basic forms of deep foundations are possible, and many special ones are available as priority systems. Individual systems become popular regionally due to a combination of availability, appropriateness to local problems, and marketing concentration by companies.

Piles

Piles are elements driven into the ground in the manner of large nails. Various special means of advancing piles are possible, including vibration, but the usual means is simply to smash them in. The two basic types are:

Friction piles: These simply develop resistance to being further pushed into the soil due to friction on their surfaces, directly analogous to a nail in wood. Their capacity is inferred by the difficulty encountered in advancing them the last few feet.

End-bearing piles: These are driven until their ends encounter extreme resistance—usually because of rock, but sometimes simply a very hard layer of soil beneath much softer ones.

Piles take various forms and use various combinations of wood, steel, and concrete. Some common examples are shown in Figure 4.9. Various priority systems consist of the driving of a steel shell, using a solid steel core (called a mandrel) during the driving. The mandrel is withdrawn when resistance is adequate and the hollow shell is

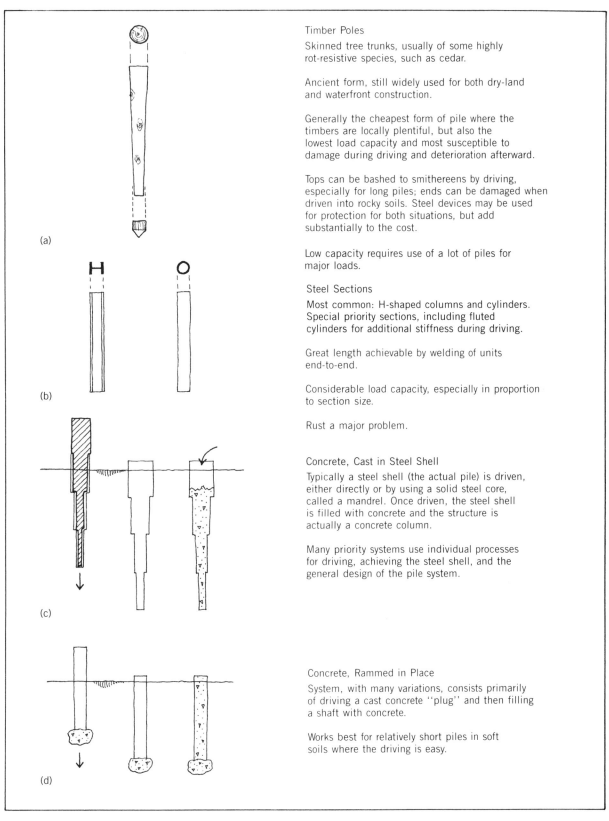

Timber Poles

Skinned tree trunks, usually of some highly rot-resistive species, such as cedar.

Ancient form, still widely used for both dry-land and waterfront construction.

Generally the cheapest form of pile where the timbers are locally plentiful, but also the lowest load capacity and most susceptible to damage during driving and deterioration afterward.

Tops can be bashed to smithereens by driving, especially for long piles; ends can be damaged when driven into rocky soils. Steel devices may be used for protection for both situations, but add substantially to the cost.

Low capacity requires use of a lot of piles for major loads.

Steel Sections

Most common: H-shaped columns and cylinders. Special priority sections, including fluted cylinders for additional stiffness during driving.

Great length achievable by welding of units end-to-end.

Considerable load capacity, especially in proportion to section size.

Rust a major problem.

Concrete, Cast in Steel Shell

Typically a steel shell (the actual pile) is driven, either directly or by using a solid steel core, called a mandrel. Once driven, the steel shell is filled with concrete and the structure is actually a concrete column.

Many priority systems use individual processes for driving, achieving the steel shell, and the general design of the pile system.

Concrete, Rammed in Place

System, with many variations, consists primarily of driving a cast concrete "plug" and then filling a shaft with concrete.

Works best for relatively short piles in soft soils where the driving is easy.

Figure 4.9 Forms of piles: (a) Timber poles. (b) Steel sections. (c) Concrete, cast in steel shell. (d) Concrete, rammed and packed on top of a driven concrete plug.

filled with concrete. In some soils, the shell can also be withdrawn during the placing of the concrete. In other systems the shell itself is driven, usually for development of an end-bearing condition.

The oldest form of pile, and still one of the cheapest, is the simple timber pole. If rot or consumption by living organisms can be handled, these are still widely used. For permanent, major construction, however, steel and concrete are now favored—sometimes simply because of their larger load capacities.

Piers or Caissons

These are literally tall columns of concrete, constructed in an excavated shaft. They function in a way similar to end-bearing piles, with ends inserted into rock (called socketed ends) or widened to bear on hard soil (called belled ends). They may range in size from small (18 in. diameter) up to gigantic—the latter for high-rise buildings, large bridge piers, and so on.

Where soil conditions permit, piers of small to moderate size and relatively short length may be excavated by drilling, in the general manner used for postholes or water wells. Large piers, however, must simply be dug out, with lining of the shaft walls installed as the hole is advanced.

The term *caisson*, which is still commonly used to describe this element, actually comes from a method that was developed many years ago for advancing excavations for large piers in soft soils. This consisted of building a working chamber (the *caisson*—French for "box") with no bottom and then digging out the soil to steadily lower the chamber. Once lowered to its desired location, the chamber becomes the bottom of the pier. This is still used, but mostly for bridge piers under water.

Figure 4.10 shows the general form of piers used for building construction.

Choice of foundation systems involves many considerations, including factors relating to soil conditions, general excavation requirements, size and type of the building project, local availability of foundation subcontracting work, and experiences with other recent construction. A lot of advice from qualified consultants is indicated if any serious problems are anticipated.

Construction on Piles and Piers

The general planning of buildings that are supported on piles or piers, rather than simple bearing footings, requires some special considerations. These may differ for various particular forms of the deep foundation elements, but have some typical conditions, including the following.

Point supports: Piers or clusters of piles must be spaced some distance apart. This does not generally affect planning of columns, but does cause differences for walls and slabs on grade.

Minimum load unit: Footings can be made quite small, appropriate to their required load-carrying tasks. Piles and piers typically have a minimum size which may represent a significant load capacity. This is even more critical for column loads on piles, since the minimum pile cluster is usually one with three piles, and is especially critical for one-story buildings where column loads are mostly from light roofs.

Use of heavy equipment: Installation of piles and drilled piers requires heavy equipment, which cannot economically be transported great distances for small

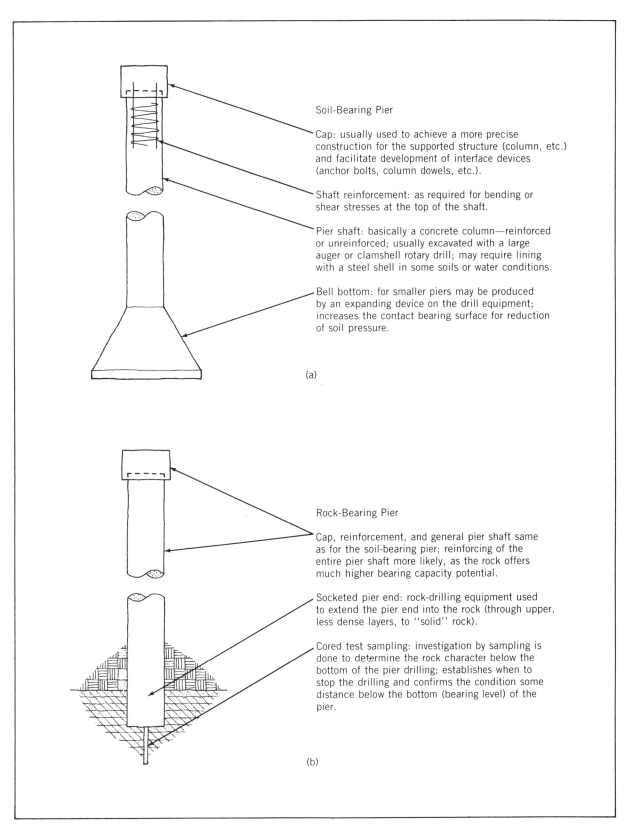

Soil-Bearing Pier

Cap: usually used to achieve a more precise construction for the supported structure (column, etc.) and facilitate development of interface devices (anchor bolts, column dowels, etc.).

Shaft reinforcement: as required for bending or shear stresses at the top of the shaft.

Pier shaft: basically a concrete column—reinforced or unreinforced; usually excavated with a large auger or clamshell rotary drill; may require lining with a steel shell in some soils or water conditions.

Bell bottom: for smaller piers may be produced by an expanding device on the drill equipment; increases the contact bearing surface for reduction of soil pressure.

(a)

Rock-Bearing Pier

Cap, reinforcement, and general pier shaft same as for the soil-bearing pier; reinforcing of the entire pier shaft more likely, as the rock offers much higher bearing capacity potential.

Socketed pier end: rock-drilling equipment used to extend the pier end into the rock (through upper, less dense layers, to "solid" rock).

Cored test sampling: investigation by sampling is done to determine the rock character below the bottom of the pier drilling; establishes when to stop the drilling and confirms the condition some distance below the bottom (bearing level) of the pier.

(b)

Figure 4.10   Forms of excavated piers: (a) With widened bottom for bearing on soil. (b) With bottom seated into rock.

projects. Access to difficult sites (hillsides, swamps, islands, etc.) may be a problem. Operation of pile drivers may upset the neighbors.

For any project, it is highly desirable to anticipate the form of foundation system very early in the planning stages for both buildings and general site development. This relates to both the economic feasibility of the construction and the intelligent planning of the work.

Information Sources

Simplified Design of Building Foundations, 2nd ed., James Ambrose, New York: Wiley, 1988.

Sweets Catalog Files, for priority foundation systems.

4.6 SOIL-RETAINING STRUCTURES

Site development frequently involves the use of various retaining structures. These help to achieve abrupt changes in the ground surface elevation. The form of construction often relates primarily to the height of the elevation change on the two sides of the retaining structure.

Curbs

The smallest retaining structures are curbs, which may take various forms, as shown in Figure 4.11. Curbs are usually edging devices that define the boundary between different units of the site surface. They are frequently placed at the edges of pavements and thus often relate to the paving materials and forms. The form of drainage of the pavement may also affect the details of the curb.

Curbs are usually limited to achieving elevation changes of less than 18 in. or so. As the height of the retaining structure increases, a different general form of construction is required.

Loose-Laid Retaining Walls

For abrupt elevation changes of more than 18 in. or so, some form of wall construction is required. This may be achieved with loosed-laid stones (without mortar) or other elements, as shown in Figure 4.12. Such construction must be banked, or leaned, into the cut to resist the horizontal force of the soil on the high side of the wall.

Walls of this type can be very effective and simple to construct. Executed with fieldstone or timber, they may be quite attractive, especially when used with other largely natural materials, such as earthen surfaces or plantings.

A typical advantage of the loose-laid wall is its natural porosity, which allows groundwater to seep through. This is especially critical when regular heavy irrigation of plantings on the high side occurs.

Cantilever Retaining Walls

The strongest retaining structures for achieving abrupt elevation changes are *cantilever retaining walls,* consisting of structural walls of masonry or concrete anchored to a large footing. Some common forms are shown in Figure 4.13.

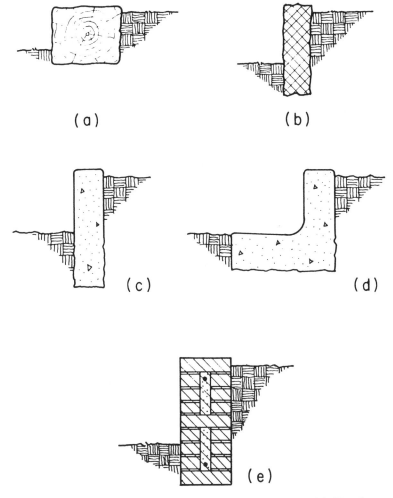

Figure 4.11   Forms of curbs: (a) Timber or log. (b) Stone. (c) Simple concrete. (d) L-shaped concrete with gutter. (e) Brick.

A form used for walls up to around six feet high is that shown in Figure 4.13(a). (Height refers to the ground surface elevation difference on the two sides of the wall.) Critical structural concerns are for the rotational, overturning effect and the horizontal sliding effect, both caused by the horizontal soil pressure on the high side of the wall. The dropped portion of the footing (called a shear key) is commonly used to enhance the resistance to horizontal sliding.

Short walls may be achieved with masonry or concrete. If achieved with concrete, both the wall and the footing are usually reinforced as shown in Figure 4.13(a). Masonry walls may be constructed similarly—as shown in Figure 4.13(b)—or may be developed as gravity walls, relying strictly on their dead weight to resist the overturning effects, as shown in Figure 4.13(c).

Gravity retaining walls are still frequently used where large units of stone are available. The all brick construction shown in Figure 4.13c is an ancient form, but probably not applicable in most situations today, unless a mass of unmatched recycled bricks are available. The gravity wall of stone may generally resemble the form of the loose-laid wall in Figure 4.12a, except for the addition of mortar between the units. Another form for the gravity wall is that using large stones cast into a concrete wall, as shown in the photograph in Figure 3.4b and as illustrated for Building 5 in Chapter 7 (see Figure 7.22).

As walls get taller, it is common to use a tapered wall form. Concrete walls are evenly tapered, as shown in Figure 4.13(d), while masonry walls are typically step-tapered, using regular units of the masonry, as shown in Figure 4.13(e).

Tall retaining walls may also be braced by buttresses or fin walls perpendicular to the retaining wall, as shown in Figure 4.13(f) and (g). If built on the back side, these do not affect the viewed side of the wall, although building them on the exposed, low side is usually easier and more economical, as it involved less excavation and backfill on the high side.

Retaining walls may also be developed as parts of building construction, becoming building walls with some bracing of the wall often provided by other elements of the

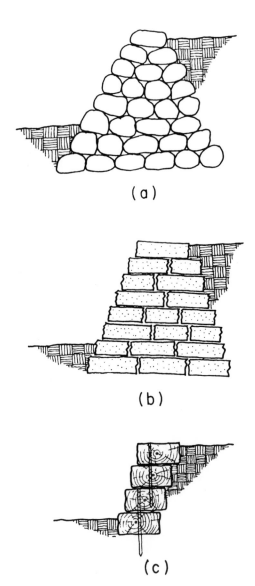

(a)

(b)

(c)

Figure 4.12 Loose-laid retaining walls: (a) With stones. (b) With broken concrete slab pieces. (c) With timbers (railroad ties).

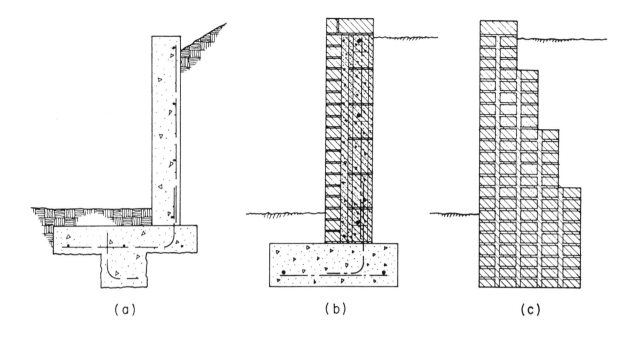

(a)   (b)   (c)

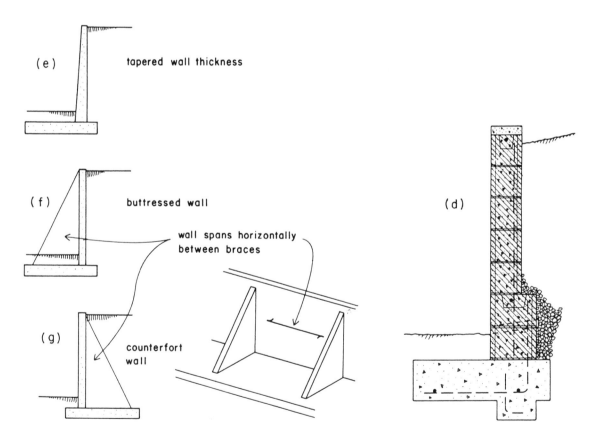

(e)   tapered wall thickness

(f)   buttressed wall

wall spans horizontally
between braces

(g)   counterfort
       wall

(d)

Figure 4.13 Forms of retaining walls: (a) Short reinforced concrete cantilever wall. (b) Reinforced concrete block wall. (c) Brick wall as a gravity wall, relying on its mass for stability. (d) Tall concrete block wall with tapered form. (e) Tall concrete wall, tapered. (f) Tall buttressed wall. (g) Tall counterfort wall.

building construction. The typical basement wall is generally a retaining structure, although not usually of the cantilever type, as it works by spanning between the basement and first-floor constructions which brace it laterally.

Temporary Bracing

Temporary bracing structures of various types are frequently used during construction. Various types of structures and the situations in which they are used are discussed in Section 4.8. These structures are often used to brace cuts so that excavations can be achieved without collapse of the cut face into the excavation. In many cases the construction (basement wall, cantilever retaining wall, etc.) may serve as a permanent retaining structure to replace the temporary one.

Information Source

Simplified Design of Building Foundations, 2nd ed., James Ambrose, New York: Wiley, 1988.

4.7 PAVEMENTS

Paved surfaces may be achieved with various materials, common ones being the following:

Concrete: This is usually the strongest form of paving and can be used for heavy-traffic-bearing roads or simple walks. Concrete paving slabs are typically placed over a thin layer of gravel or crushed rock to provide a good base and a draining layer beneath the pavement. Minimal reinforcement may be provided with a single layer of steel wire fabric; thicker pavements may be reinforced with two-way grids of steel rods.

Asphalt: Various forms of concrete produced with an oil-based binder (tar, etc.) can be used. Materials, thickness, and base preparation depend on the desired degree of permanence and limits on cost.

Loose-unit pavers: Bricks, cobblestones, cast concrete units, cut wood sections, or other elements may be laid over a sand and gravel base. This is an ancient form of paving and can be very durable if installed properly and made with durable elements.

Loose materials: Fine gravel or pulverized bark may be used for walks or areas with light traffic. These generally require some ongoing maintenance to preserve the surface, but can be very practical and blend well with natural features of a site (existing surfaces, plantings, etc.).

Areas to be paved must be graded (recontoured) to some level below the desired finished surface to allow for the installation of the paving. If existing site materials are undesirable for the pavement base, it may be necessary to cut down to a lower surface elevation and to import materials to build up a better base for the pavement.

Pavements—especially of the solid form of concrete or asphalt—result in considerable surface runoff during rainfall, which must be carefully considered in the general investigation of site surface drainage.

4.7.1 Concrete Slabs

The usual means of achieving a working surface for sidewalks, driveways, basement floors, and floors for buildings without basements is to cast concrete directly over some prepared base on top of the soil. As shown in Figure 4.14, the typical construction for this consists of the following components:

1. A prepared soil surface, graded to the desired level; compacted, if necessary to avoid subsidence.
2. A coarse-grained pavement base, usually predominantly fine gravel and coarse sand; also compacted to some degree, if more than a few inches thick. Three or four inches is common for floor slabs.
3. A membrane of reinforced, water-resistant paper or thick plastic film (6 mil minimum), used where moisture intrusion is especially severe.
4. The concrete slab.
5. Steel reinforcement; often of heavy-gage wire mesh.

While the basic construction process is quite simple, there are various considerations that must be addressed in development of details and specifications for the construction.

Thickness of Pavements

For residences and other situations with light traffic, a common thickness is a nominal 4-in. slab; actually 3.5 in., if standard lumber 2 × 4's are used as edge forms. With proper reinforcement and good concrete, this is an adequate slab for most purposes.

Where some vehicular traffic is anticipated, or where other heavy concentrated loads—such as those from tall, heavy partitions—are expected, it may be desirable to jump to the next logical thickness of 5.5 in., based on using 2 × 6's for edge forms. At this thickness, it may be possible to avoid providing individual wall footings for nonstructural partitions, a simplification in construction probably justifying the cost of the additional concrete for the slabs.

For very heavy trucks, for storage warehouses, or for other situations involving heavy concentrated loads, thicker pavements may be necessary. However, the nominal 4-in. and 6-in. slabs account for most building floors.

Reinforcing

For 4-in. slabs reinforcing is usually accomplished with welded wire mesh. For thicker slabs, or as an alternative for 4-in. slabs, small-diameter rebars may be used with somewhat wider spacings than those in the wire mesh. Even when mesh is used, some extra rebars may be provided at edges, around openings, or at other critical locations.

The objective for the reinforcement is to reduce cracking of the concrete, due primarily to shrinkage during curing, differential volume changes due to temperature changes, and some unequal settlement of the pavement base. Cracking is especially undesirable in the top (exposed surface) of the slabs, so the reinforcement should be held up during casting to be relatively close to the top surface.

Another means for reinforcing—or basically altering—the concrete is to add fiber materials to the mix, a growing practice for all exposed slabs. In cases where considerable movement of the base is expected, pavements may need to be prestressed or developed as framed systems on grade.

Joints

It is generally desirable to pour paving slabs in small units, primarily to control shrinkage cracking. For floor slabs of buildings, this results in construction joints, which may not relate well to development of floor finish materials. Where permanent wall construction is established, joints should be located at these points, rather than in the middle of floor spaces.

For exterior paving, joints should occur quite frequently, appropriate to the pavement width, thickness, and general desire to be able to replace settled or cracked units easily. Joints should also be placed at what are predictable locations for cracks, such as changes in width, intersections, and corners. Extreme outdoor temperature range or anticipated soil settlements may also affect the frequency of joints.

Larger pours can be made if control joints are formed or cut in the slab. Movements at these joints may be small for interior floors, but they should still be located at points that will not cause problems with floor finishes if possible.

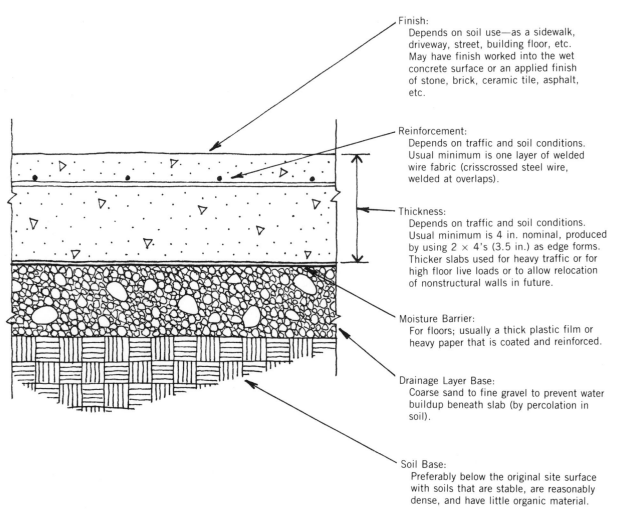

Finish:
Depends on soil use—as a sidewalk, driveway, street, building floor, etc. May have finish worked into the wet concrete surface or an applied finish of stone, brick, ceramic tile, asphalt, etc.

Reinforcement:
Depends on traffic and soil conditions. Usual minimum is one layer of welded wire fabric (crisscrossed steel wire, welded at overlaps).

Thickness:
Depends on traffic and soil conditions. Usual minimum is 4 in. nominal, produced by using 2 × 4's (3.5 in.) as edge forms. Thicker slabs used for heavy traffic or for high floor live loads or to allow relocation of nonstructural walls in future.

Moisture Barrier:
For floors; usually a thick plastic film or heavy paper that is coated and reinforced.

Drainage Layer Base:
Coarse sand to fine gravel to prevent water buildup beneath slab (by percolation in soil).

Soil Base:
Preferably below the original site surface with soils that are stable, are reasonably dense, and have little organic material.

Figure 4.14   Typical sitecast concrete pavement (slab on grade).

Surface Treatment

If the top of a paving slab is to be exposed to wear, it should be specially formed for this purpose during casting of the concrete. A hard, smooth surface is generally desired, typically developed with fine trowelling with a steel trowel, and possibly the addition of some hardening materials to the finished surface.

For a slip-resistant surface, intended to reduce accidents when a pavement is wet, a grit material may be added to the surface or a deliberately roughened surface developed (broomed, etc.). Smoothness is also generally not highly desirable when other finish materials are to be used, especially ones that must be attached with adhesives. In the latter case, the steel troweling may be omitted and a rough leveling of the surface acceptable.

4.7.2 Other Paving Materials

Concrete pavements are highly durable and effective for heavy traffic. They are, however, quite expensive if correctly installed, so other forms may be used for economy or because of a desire for a different form of surface. Of course, most of these can be developed on top of a concrete slab—as they might be when used on the roof of an underground structure. The following discussion, however, deals primarily with their use in place of a concrete slab. (see Figure 4.15.)

Asphalt

Asphalt paving is essentially a form of concrete (sand, gravel, and a binder) made with a bituminous binder instead of portland cement. However, it is typically a built-up layer, with a somewhat denser, smoother material developed for the top surface. It is naturally less permeable (allowing water penetration) than ordinary concrete, although rapid water accumulation during heavy rains will flow off a sloping surface of either concrete or asphalt at about the same rate.

As with a concrete slab, a base should be developed for a good asphalt pavement. However, for lightly trod walks, a very thin coating of the ground may suffice. Asphalt surfaces are somewhat less durable than hardened concrete, and constant traffic will wear them down. However, they can be repaired or restored by additional coatings on top of the old paving.

Unit Pavers

Ancient pavings were developed with stones laid to produce a relatively smooth surface. Very fine surfaces were developed with cut stones, and the floors over soil in many great works of architecture were developed as loose stone paving over a crushed gravel base.

Later, bricks and fired clay tiles were used in the same manner as cut stones. Matched sets of smooth fieldstones were used to produce cobblestone paving; these stones were basically considered peasant-quality stuff—hence the word *cobble,* which meant roughly formed.

As with any paving of loose units, the application may be made essentially directly over soil or developed more permanently. Bricks or tile, for example, may be set in mortar over a concrete slab for a really permanent, stable pavement.

In addition to stone, bricks, and tile, just about any hard elements can be used for

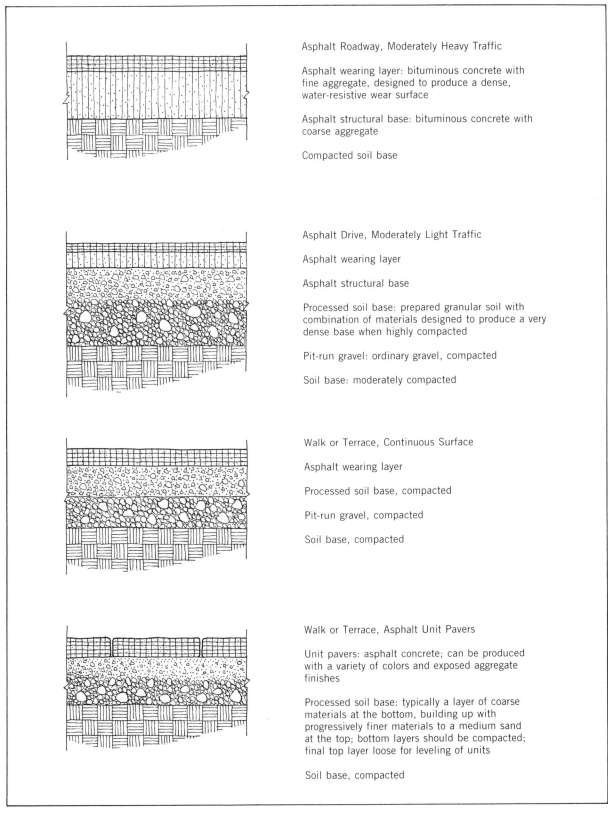

Asphalt Roadway, Moderately Heavy Traffic

Asphalt wearing layer: bituminous concrete with fine aggregate, designed to produce a dense, water-resistive wear surface

Asphalt structural base: bituminous concrete with coarse aggregate

Compacted soil base

Asphalt Drive, Moderately Light Traffic

Asphalt wearing layer

Asphalt structural base

Processed soil base: prepared granular soil with combination of materials designed to produce a very dense base when highly compacted

Pit-run gravel: ordinary gravel, compacted

Soil base: moderately compacted

Walk or Terrace, Continuous Surface

Asphalt wearing layer

Processed soil base, compacted

Pit-run gravel, compacted

Soil base, compacted

Walk or Terrace, Asphalt Unit Pavers

Unit pavers: asphalt concrete; can be produced with a variety of colors and exposed aggregate finishes

Processed soil base: typically a layer of coarse materials at the bottom, building up with progressively finer materials to a medium sand at the top; bottom layers should be compacted; final top layer loose for leveling of units

Soil base, compacted

Figure 4.15 Miscellaneous forms of paving.

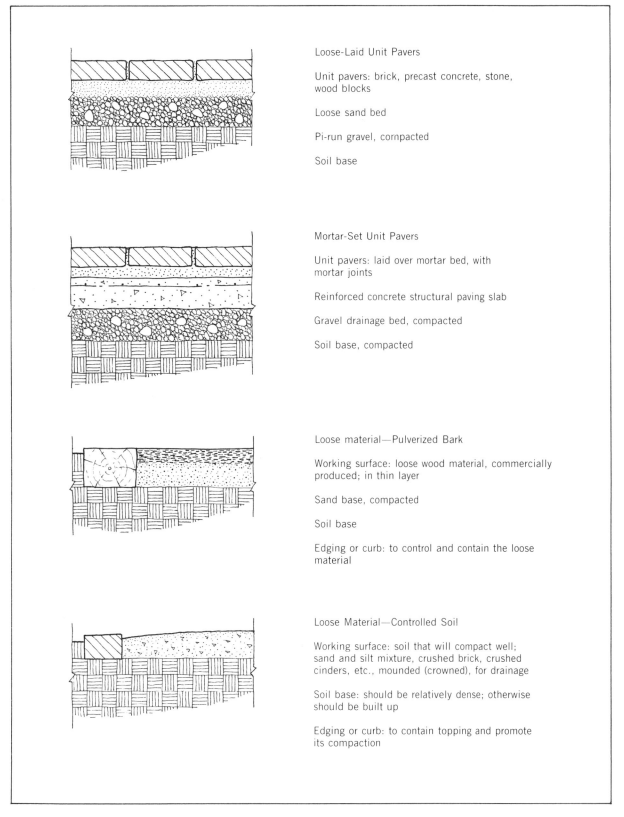

Loose-Laid Unit Pavers

Unit pavers: brick, precast concrete, stone, wood blocks

Loose sand bed

Pi-run gravel, compacted

Soil base

Mortar-Set Unit Pavers

Unit pavers: laid over mortar bed, with mortar joints

Reinforced concrete structural paving slab

Gravel drainage bed, compacted

Soil base, compacted

Loose material—Pulverized Bark

Working surface: loose wood material, commercially produced; in thin layer

Sand base, compacted

Soil base

Edging or curb: to control and contain the loose material

Loose Material—Controlled Soil

Working surface: soil that will compact well; sand and silt mixture, crushed brick, crushed cinders, etc., mounded (crowned), for drainage

Soil base: should be relatively dense; otherwise should be built up

Edging or curb: to contain topping and promote its compaction

Figure 4.15 *(continued)*

paving. Wood blocks, planks, timbers, or cut sections of logs can be used. So can precast concrete units—now most commonly used when installation is directly in the ground.

Complete details for unit paving will vary depending on the units themselves, but also on the traffic to be borne and what is underneath the paving. Selection of the units is usually a landscape design decision, but also responds to availability of materials and general economics for the project.

Loose Materials

For some forms of traffic, or for special surfaces such as racetracks, playgrounds, or playing fields, various loose materials can be used for a developed surface. Widely spaced unit pavers may be used with an infill of grass for a form of paved lawn. Other nongrassy areas can be developed with pulverized bark, sawdust, fine gravel, or various mixtures of soil materials.

Natural soil surfaces may form relatively hard, durable crusts in some cases, or be encouraged to do so with some help. Addition of some fine silt or dry clay to an existing sand or gravel surface and use of watering or compaction may produce such a surface.

Grass alone, when carefully cultivated and maintained, can form a type of pavement. Besides sustaining some degree of traffic, it may serve effectively to prevent surface erosion, improve precipitation runoff, or otherwise do what is generally expected from other forms of paving.

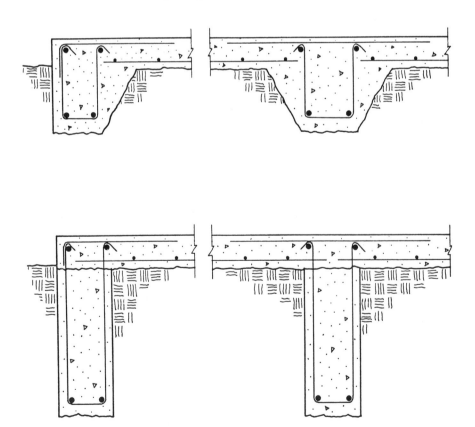

Figure 4.16 Framed slabs on grade: Upper—single pour. Lower—with separately formed and cast beams.

4.7.3 Framed Pavements

In some situations, it is necessary to develop a paved surface for which the stability of the supporting ground cannot be firmly established. One common situation of this kind is that which occurs when a building floor slab must be placed on considerable fill. Foundations for the building columns and walls will hopefully be supported with more stability, so the problem becomes one of relative settlement of the floor with respect to other construction.

Of course, methods exist for consolidation of fill materials, but they are mostly quite labor intensive and must be achieved in relatively thin, successive layers. The greater the total vertical thickness of the fill, the less feasible compaction becomes.

One method of dealing with this problem is simply to develop the pavement (usually a concrete slab) as a spanning, reinforced concrete structure. Possible forms for this are shown in Figure 4.16. For relatively short spans, beams may be formed as simple trenches and cast as one with the slab (Figure 4.16(a)). For larger spans, or when considerable fill is placed (often the major reason for developing the structure), it may be more feasible to form and cast beams first, as shown in Figure 4.16(b). Examples of these floor systems are shown in Chapter 7 for buildings 1 and 6.

Light partition walls may usually be supported on concrete floor slabs on grade without special provision. Walls used as bearing walls, or simply any walls of heavy construction, should have some more developed support—essentially a wall footing. The construction shown in Figure 4.16 may also be used for this purpose in some situations.

4.8 DEVELOPMENT OF GROUND SURFACE FORM

Development of sites often involves some considerations for the maintenance or manipulation of the form of the ground surface. This section treats some of the common issues and situations encountered in site development.

4.8.1 Slope Control

Site development and building construction often require the treatment of sloping ground surfaces. When this is necessary, a critical decision is that regarding the maximum feasible angle for the slope (see Figure 4.17(a)). The stability of a slope may be in question for a number of reasons. Two principal issues relating to the soil materials are the potential for erosion from excessive rainfall or irrigation and the possibility for movement of the soil mass in a downhill direction.

A sloped surface may be generated in two different ways: by cutting back of existing soil, or by building up with fill. The relative stability of the slope will depend largely on the character of the soils at and near the surface of the slope.

For cut slopes the soil is largely existing and slope angle limits must be derived basically from the determination of existing soil conditions. For a relatively clean sand the angle of repose is quite easily derived, although erosion or shock may be critical, particularly for loose sands. For rock or some very stable cemented soils, a vertical cut face—or even a cave—may be possible.

For most ordinary surface soils of a mixed character, the proper slope angle must be derived from studies of potential loss of the slope face by various mechanisms of failure. Slope loss by either erosion or slippage is often due to a combination of water soaking of the surface soils and the presence of a relatively loose soil mass.

Erosion is a problem related to many considerations of the general site development. The slope itself or the water it must drain may be affected by plantings, irrigation, general site contouring, pavements, and both site and building construction. Slopes may be surfaced with plantings or various pavings to protect the surface and retard erosion.

Slippage of the soil mass on the slope may occur from a number of causes. A common one is simply the vertical movement of the soil mass due to gravity, as shown in Figure 4.17(b). This may occur as a semirotational effect along a slip plane, as shown, or by the sliding of one soil mass over another, as shown in Figure 4.17(c).

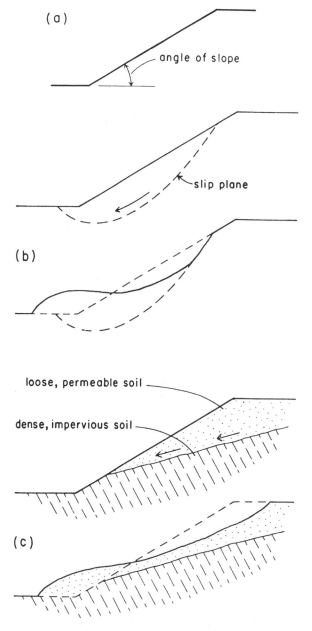

Figure 4.17 Considerations for slope failures: (a) Definition of the slope angle. (b) Rotational slope plane failure in loose soils. (c) Slippage of layers in a stratified soil mass.

For many ordinary situations reasonably safe slope angle limits are established by experience and rules of thumb. Conservative limits may be established by building codes, although studies by a geotechnical expert or provision of some slope control elements may allow steeper angles.

4.8.2 Retaining Devices for Slopes

Slopes or abrupt changes in the ground surface elevation can be maintained or established by various means. Simple shaping of the soils may suffice if the mechanisms of slope failure are acknowledged, as discussed in the preceding section. However, various other elements can be used to help, as shown in Figure 4.18.

Surface Treatments

Surface treatments are basically means for keeping the surface itself in place. They do not generally affect deeper soil conditions, so they cannot overcome slip plane failure or sliding, as illustrated in Figure 4.17(b) and (c). They are generally most effective in simply retarding erosion or the massive soaking of the soil from continued precipitation. Roots of various plantings can grab a significant layer of the surface once the plantings are well established. Coatings of asphalt, individual pavers, coarse gravel, or large rocks may be used. Use of geotextiles can also be one factor in a total slope stabilization system.

Slopes may also be subdivided into stepped units. The "steps" may be quite large and individually maintained as close to flat—or at least a much shallower angle than the general slope. The "risers" of these steps can be any form of retaining structure appropriate to the height of the riser. Individual steps may be occupied by plantings, walkways, driveways or roads, or a series of buildings. Large developed hillsides are often stepped up with roads and rows of buildings in this manner.

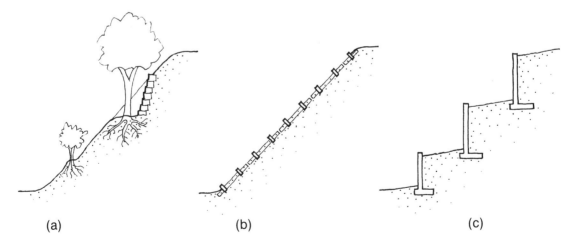

(a)   (b)   (c)

Figure 4.18   Devices for slope retention: (a) Plantings with strong root growth. (b) Coverings of paving or specially-shaped units. (c) Stepped construction with retaining structures.

4.8.3 Temporary Bracing

Various expedient methods may be used to maintain soil profiles during construction or general site development. Deep excavations for buildings or general construction on steep slopes may utilize these methods.

Shallow excavations can often be made with no provision for the bracing of the sides of the excavation. However, when the cut is quite deep, or the soil has little cohesive character (loose sand, for example), or the undercutting of adjacent construction or property is a concern, some form of bracing for the cut faces may be required.

One means for bracing consists of driving steel sheetpiling to form a wall that can then allow soil to be removed from one side, as shown in Figure 4.19(a). As with other

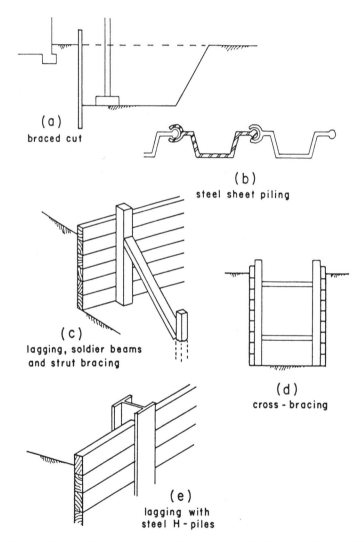

Figure 4.19   Elements used for bracing vertical cuts: (a) Driven bracing. (b) Plan section profile of steel sheet piling, used commonly as in (a). (c) Soldier beams and lagging, shown with raker braces. (d) Narrow excavation with soldier beams and lagging, braced across the hole (no rakers). (e) Lagging with steel column shapes used for soldiers; the soldiers may be driven or not.

temporary walls of this nature, the piling may stay in place, possibly used as the forming for one side of a cast concrete wall. However, if possible, the quite expensive piling units can also be withdrawn for reuse.

Another form of bracing wall is one constructed with vertical beams or posts (called soldiers) and horizontal boards or sheets (called lagging), as shown in Figure 4.19(c), (d), and (e). This construction may be achieved by driving the soldiers as piles, but it is more often done by building the wall downward as the excavation proceeds.

Both sheetpile walls and those using soldiers and lagging generally need some form of bracing themselves. This may take the form of cross-hole braces, as shown in Figure 4.19(d), or individual struts (called rakers), as shown in Figure 4.19(c). For walls intended for permanent installation, where soil conditions permit, bracing may be done through the wall and into the soil mass with drilled-in anchors.

Major bracing of excavations is frequently required with construction in tight urban site situations, especially when the site is surrounded by existing streets or buildings. Its development must be done as a coordinated effort with the general foundation and site construction work.

5

Special Concerns

This chapter treats a number of special concerns that affect the development of construction for sites and below-grade portions of buildings.

5.1 MANAGEMENT OF SITE MATERIALS

An existing site represents an inventory of materials. Decisions must be made that involve the management of this inventory in the process of site development. These decisions may involve the removal, rearrangement, modification, or replacement of materials. The nature of the existing site and the type of redevelopment will determine the extent to which site modification is required.

Site Materials

Site materials can be broadly divided between those below grade and those above. Below grade are mostly various soil, rock, and water deposits. For previously developed sites there may also be various below-grade constructions. For heavily forested sites, there may be considerable root growth below grade.

Above the ground surface may be trees or other plant growth. Preservation or removal of these must be determined in conjunction with the site development plans. This is discussed in Section 5.3. If preservation is desired, it may be necessary to develop careful plans to protect existing growth during site work and construction. Raising or lowering of the grade, major changes in surface drainage, and other modifications may seriously affect existing growth.

General concerns for soil materials are discussed in the Appendix, which also describes various soil properties and the means for identification of soil types. Soil is basically natural, but modifications are possible—and indeed often necessary—for site development.

In general, existing site materials must be viewed as a given inventory of materials that must be managed in the site development process. The existing materials may be removed, replaced, modified, relocated on the site, or simply left as is.

Problems that may affect the management of site materials include the following:

Establishment of finished grades: If the level of the site surface (grade) is to be substantially raised or lowered, there will be major needs for removal or importation of materials.

Building excavation: For large buildings with extensive below-grade construction, excavation for construction may involve extensive removal of site materials. These may be used elsewhere on the site, or require considerable planning for transportation and disposal off the site.

Extensive landscaping: Major new plantings may require the importing of considerable material for surface soils to sustain plant growth.

Site construction: Extensive development of site structures may require removal of soils displaced by the construction.

Special problems involving these and other concerns are discussed in the remaining sections of this chapter. The anticipation of these problems will often affect the type and extent of data required from site surveys and subsurface investigations.

Removal of Soil

It is desirable, when possible, to balance the cuts and fills required to achieve finished grades so that no significant removal or importing of soils is required. This is not always possible, however, and major removals may sometimes be required.

A common situation requiring extensive removal of soil is that of major below-grade construction. Most sites are not raised or lowered significantly, simply because they must retain some connection to the site boundary conditions of neighboring properties, streets, or underground utilities. Extensive below-grade site construction therefore displaces soils that must be removed.

The other major reason for removal of existing site materials is that they are undesirable for some reason. For plantings, for better support of pavements, for drainage, or for various reasons, the existing soils may not be usable and not feasibly modifiable.

Removed soils must be taken somewhere, which may present a major problem to be solved for the site development, the more so if they are basically undesirable for some reason.

Imported Materials

In the best of situations, the materials desired for importing to one site may be those required to be removed from another site. Where extensive, ongoing construction occurs, this exchange is frequently made.

Topsoil for plantings required for one site may be removed from another site that is to be mostly covered with buildings, site constructions, and pavements. Or, the soil removed to achieve a major excavation on one site may be used to raise a major depressed portion of another site.

Where this is not the case, the practicality of a particular proposed site development may hinge on the feasibility of obtaining the necessary imported soil materials.

Modification of Site Materials

Existing site materials often represent usable raw materials that require some modification for the purposes of the site development. Surface soils to be used for surfacing may need to be cleaned of debris, large roots, rocks, and so on. Soils to be used for structural purposes, such as base supports for pavements or bearing supports for foundations, may need some other forms of modification. These latter modifications may seek to improve soil strength, improve resistance to settlement-producing deformations, change water-related properties, or simply improve general soil stability.

Various forms of soil modification and the means for achieving them are discussed in Section A.6 of the Appendix. Feasibility of modifications should be carefully studied before decisions are made about major removal or replacement of site materials.

5.2 PRESERVATION OF SITE MATERIALS AND ELEMENTS

For sites unspoiled by previous development, it is often desirable to preserve some features. This may be a practical matter of using resource materials, as discussed in Section 5.1. However, it may also be an issue relating to conservation of existing natural site features, including especially existing plantings.

Old, slow-growth trees can often not be replaced; either feasibly or simply biologically assurably. Figure 5.1 shows some efforts made to preserve native oak trees, each several hundred years old. This is not always successful, despite major efforts and allowances. These hardy survivors of decades of droughts and fires can sometimes succumb to simple environmental changes associated with development activities.

Existing trees—or other features, such as surface rock formations or streams—may simply be in the wrong place for logical development of the site. This may be true for either horizontal or vertical placement. Building locations and street layouts may be shifted horizontally (Figure 5.1(a)), but vertical displacement (Figure 5.1(b)) is often a more difficult problem.

Skillful landscape design and site engineering may in some situations be capable of recreating "natural" site features. This may be done by relocating existing elements (trees, rocks, diverted waterways), or by building new ones to replicate prior conditions or neighboring sites.

5.3 LANDSCAPE DEVELOPMENT

What constitutes landscape development depends on who you ask to define the landscape. Gardeners will reply with a definition related mostly to plantings and associated site constructions. Landscape architects will have a broader view, relating to the perimeter conditions and general environment of the site, the context of site and buildings, day and night conditions, site lighting, graphics, traffic patterns, and so on.

We take the broader view here and consider the landscape to be the whole site and its context. Development of the landscape thus literally involves almost *everything* related to the visible site. Even the most humble, utilitarian elements (fire hydrants, gutters, street signs) must be considered as landscape elements. However, for the issues to be discussed in this section, we will concentrate mostly on those related to plantings and to site construction specifically developed for the landscape enhancement.

Successful use of plantings requires some basic considerations, major ones being the following:

Choice: Plantings used outdoors must relate reasonably to local weather conditions. Artificial irrigation or food supplies may be provided, but frost, rainfall, and

(a)

(b)

Figure 5.1 Large existing trees incorporated into site development: (a) By planning around their locations. (b) By protecting their ground surface elevation and general root periphery.

sunlight cannot be controlled much. Plants grow, and some stages of this growth must be visualized. Many houses have been overwhelmed by fast-growing plantings used by unscrupulous developers.

Adequate soil: Grass may be established in a few inches of soil, but large shrubs and trees need a significant soil mass. This becomes critical where site excavation is difficult, good soil is in short supply, or considerable below-grade construction is anticipated.

Effects of continuous irrigation: Watering of plantings, especially in arid climates where they are not "natural" growth, can cause problems with soil stability or for construction below grade.

Effects of root growth: Large trees have large roots. If soil or groundwater conditions do not favor it, the root growth may have little downward progression. Continuous irrigation of broad plantings (lawns, etc.) may promote surface growth of tree roots, simply as the easiest effort by the tree. Roots can disrupt underground structures, piping (especially sewers or tile fields), lawns, pavements, and site structures with shallow footings.

5.4 TOXIC GROUND CONDITIONS

There are many possibilities for the existence of undesirable or even dangerous materials on the surface or beneath a site. Toxic wastes, mineral deposits, gases, or pockets of organic substances can represent very difficult situations for site development.

A particular problem experiencing relatively recent attention is that of radon gas, which occurs naturally in many locations. Need for critical attention to this problem is somewhat controversial, but serious conditions are indeed possible on a local basis. Basements are especially vulnerable, but even buildings with slab on grade floors can be affected. Planners of any new building construction should be aware of this issue and take care to determine if it is a problem at a particular site.

Methane gas is commonly developed by natural organic decomposition processes wherever large amounts of organic materials are deposited. This happens with the buildup of decayed plant materials in areas with heavy natural growth as well as at dump sites. Reclaimed sites over fill often have problems with methane, and planning should anticipate the future development of gas where this is a potential of significant magnitude.

5.5 SITE/BUILDING SYMBIOSIS

A lot of buildings simply squat on their sites in a situation that is, at best, mutually tolerated by the site and the building. The building can imagine better places to park but is stuck here, while the site was most likely happier in its undisturbed state. The more character the site has, the worse the situation in some cases. Housing developments often clear away ground slopes, plantings, and rock outcroppings and redirect or eliminate natural streams and drainage channels to create boring sites.

This mismatch of site and building can become truly grotesque when a building is designed for a flat site and then placed on one with a marked nonflat nature. It is somehow weird to see a typical one-level house parked on a hillside by creating a flat site or by using stick supports to prop it up.

This is probably no more stupid then having Cape Cod cottages in the desert or

Spanish colonial adobe houses in Maine, but so many opportunities exist to utilize a site and to form a building in response to site situations that it is wasteful not to consider using them.

The building is usually the major structural element on a site, and when some special site problem exists, it may be possible to use the building as a site structure. Figure 5.2 shows a building built on a very steep slope. In this case the slope was itself already a problem, being subject to progressive erosion and landslides, presenting an ever-increasing threat to the buildings above. Retention of the slope by direct means was generally not feasible. The building shown was developed to form a gigantic retaining wall, simultaneously performing a community service as well as developing a buildable site from a seemingly hopeless one.

Another situation is shown in Figure 5.3, where a building was built in the form of a bridge across a ravine-shaped site. In this case the basic site form was retained, which might have been done to protect a traffic condition, a site drainage condition, or simply the desirable natural site form.

Another hillside building is shown in Figure 5.4—in this case a solitary building, but one highly visible from below. Partly to utilize earth sheltering for energy savings, but mostly to create structure parking that is out of view, the roof is used for parking and the building is built partly into the hillside. The building thus fits rather snugly on

Figure 5.2   Building placed on what is usually considered an unbuildable site due to the high slope and possible undermining of upper construction. In this case, the building actually constitutes a very high retaining structure, restoring the stability of the slope and actually protecting the upper properties.

Figure 5.3   Building constructed to span a site, preserving the general site form. Pasadena Art Center, Pasadena, California. Craig Ellwood, architect. (a) View at building entry level. (b) View of entry road and spanning portion of building.

(a)

(b)

Figure 5.4   Building on a high slope, built partly into the slope and developed with roof parking that is generally out of view from the community below. City Hall, Thousand Oaks, California. Architects: Albert C. Martin Associates, Los Angeles. (a) View from below. (b) View from uphill position; showing the rooftop parking.

the hill, making for a considerably less commanding appearance. It is thus highly visible, but not seemingly overpowering: an appropriate demeanor for a local government building. And, it neither denies nor destroys the hill itself.

Finally, a different situation is shown by the building in Figure 5.5. In this case the building is built directly over a permanent deep excavation into a tar pit that yields continuing finds of prehistoric animal remains. Besides protecting the excavation, the building provides laboratory facilities, specimen storage, and a public museum. It is also a major theme structure for a fully developed park site.

The examples shown here are all buildings that required some exceptional consideration for the building/site interaction. Most buildings do not have such special concerns, but almost all buildings can be more fully developed with some careful

Figure 5.5   Building partly below ground. A major portion of this building is below the ground, extending into a deep paleontological dig and providing permanent facilities to support and display the work. Page Museum at the La Brea Tar Pits, Los Angeles, California. Architects: Thornton, Fagan, and Associates, Pasadena, California.

study of the improvement of the building/site relationships. This is, of course, only the first step in developing a full building/environment relationship that may well extend beyond the site boundaries.

5.6   WATER CONTROL

Control of water presents major problems in the development of building construction. A major part of all liability cases filed by building owners against their architects and builders have to do with water problems: leaking roofs, walls, and basements, and various site failures. Specific concerns for sites and below-grade structures are discussed throughout this book. This section presents some discussion of the overall issues of water control.

Surface Drainage

Every site must be designed for some magnitude of water that it may receive from rainstorms or melting of snow and ice. Rain is the usual problem, although downslope sites in mountainous areas can receive massive flooding from melting snow. Rainstorms can be disastrous for the rate of water delivered in a short deluge or for the total cumulative amount from an enduring rain. In some areas, rain may occur so frequently that flooding is built up slowly but steadily.

Design for surface drainage begins with a careful analysis of the predictable storm conditions and some real data for accommodation of water flow on the site. This flow must be handled on the site itself, but the real problem is often how to get it off the site and what form of disposal is available. Delivery to existing storm sewers or into streets with curbside gutters is relatively easy; simply dumping it on the neighbors may present some problems.

Surfaces to be drained may include roofs, paved areas, planted areas, and open, undeveloped ground surfaces. Each surface has particular characteristics for partially absorbing the water or accommodating a certain flow rate, depending on the slope and surface materials of the area. Effects on surfaces include the possible leaking of roofs, washing out of plantings, and general erosion of soil and loose paving materials.

The general means for controlling water flow on the site surface include the contouring (shaping) of the surface, the selection of the surface materials, and possibly the providing of some channeling devices. For very flat surfaces, such as athletic fields or parade grounds, some form of subsurface drainage may be used to reduce the buildup of water that soaks down through the surface.

In most cases, drained surface water is eventually channeled to some points for collection and disposal. There are many possibilities for this in theory, but the site itself—and neighboring property situations—will determine some limits on possible solutions. A major simple fact to keep in mind is that gravity must be acknowledged. Water can be blown by the wind or pumped uphill, but the major action is simple downhill, gravity flow.

Erosion

It is natural for surface materials from hills and mountainsides to slowly wash downhill, filling the valleys and clogging the drainage channels. If you want to keep your site surface materials, especially on steep slopes, you must make some effort to impede this natural action.

The two principal means for preventing erosion are to use ground-covering plantings or pavings. Sloped surfaces may also be covered with large stones or some form of open paving, called *rip-rap*. Use is now made of various geotechnical fabrics for this purpose, especially for protecting new plantings until they have achieved sufficient root growth.

While you don't want to lose the surface materials, it is also quite likely that the neighbors don't want to receive them. Control of erosion must be integrated with the general development of site drainage to assure that the channeling of drained materials (water plus some trash and dirt and minor surface losses) does not seriously affect the neighbors. It is likely that some site construction may be required to achieve this.

Erosion of a massive sort can occur in the form of slope failures, consisting of the sudden collapse of a slope face. This may occur due to progressive natural causes or because a slope is built up at too steep an angle for the soil conditions. The general issue of slope control is discussed in Section 4.8.1.

Protection of Construction

Water may affect site construction in a number of ways. Serious erosion can possibly remove the supporting materials or protective fills and undermine pavements or footings. Design of structures and protection of drained slopes may be affected by these concerns.

Water soaking down can also alter soil properties. This is mostly a problem with sites where the grade level is lowered significantly, bringing the effects of heavy rains into soil layers previously protected by those above them. High-void soils with cemented materials are subject to slow consolidation or even sudden collapse from this phenomenon. See the discussion in the Appendix of high-void soils.

Irrigation can also induce the preceding phenomenon, being especially of concern in arid regions where the soils have not been so affected by rain. It is probably best to simply avoid the use of heavy irrigation in this circumstance, although solutions— expensive ones—are possible. Flooding of the site before construction may achieve significant preconsolidation, or injecting of materials to fill the soil voids may prop up the soil sufficiently.

Frozen Ground

In cold climates the upper soil mass freezes during the colder seasons. Because water expands slightly upon freezing (about 4% volumetrically), this causes the ground surface to swell. Upon thawing, the swelling subsides and the surface goes down. This effect does not always occur uniformly due to the nonhomogeneous nature of soils, so some uneven distortion of the ground surface soil masses is to be expected.

This movement of the ground can have disastrous effects on pavements and on the footings for small structures. Pavements must generally be designed to ride along with the movements, but footings are usually placed at sufficient depth to be below the major frost action (usually reasonably predictable on a local basis); otherwise the supported structure will be moved. Freestanding walls, curbs, planters, and retaining walls present the most challenge in this regard.

The freeze-thaw cycles, repeated over many seasons and to some extent during cycles of warmer and colder weather in a single season, also tend to generally soften up and loosen up the top soil layers. This is great for preparing the ground for spring planting, but will eventually considerably roughen the surface of lawns or other unpaved areas of sites.

Frost effects can be reduced to some degree by measures taken to reduce the amount of water retained in surface soils. This is generally a concern at the edges of buildings, where normal provisions made to reduce water intrusion into basements may provide this protection for frost effects as well. Use of fast-draining surface soils, subsurface drains, and general shaping of the ground surface for drainage to avoid ponding effects may keep moisture levels from building up at the surface. In general, the drier the surface soils, the less the frost swelling effects.

Water Intrusion

Basements and underground spaces have major concerns for water intrusion. There are three sources of intrusion that must be considered.

Surface water: This is water from rain, from melting ice and snow, and from irrigation. If rapid surface runoff does not occur, this water may trickle down through surface soils to affect basement walls or the roofs over underground spaces.

Ground moisture: Except in extremely arid regions, some moisture exists in almost all soils. Thus contact with soil means contact with a continuously moist material, from which water can be extracted by absorption or capillary action. The concrete or masonry structures normally used for below-grade construction are quite absorptive, so some form of moisture protection (normally some surface coating) must be provided. However, other measures are also taken, such as sloping ground surfaces away from building edges, draining fill materials at the face of underground walls, and using both fast-draining soil bases and moisture barriers under concrete floor slabs on grade.

Groundwater: At points below the *groundwater level,* also called the *water table,* there is in effect standing water in the soil, capable of exerting actual water pressure against enclosures. Interior spaces extending below this level (or even close to its approximate location) must be dealt with like underwater construction. What is required here is not just *dampproofing,* as is used for resistance to moisture, but real *waterproofing,* which is like that used on flat roofs. However, while flat roofs must be designed for the presence of water, spaces under water must resist actual water pressure, so the problem is considerably more critical.

In addition, while roofs can be—and frequently are—repaired for leaks, repairing underground leaks may be quite expensive or even next to impossible. This work should be done 100% right the first time, preferably by experts.

6

Systems

System is a much overused word, but generally it describes any group of component elements that exist in some ordered relationship. Most often the group is composed to relate to some specific task. A building as a whole entity can be described as a system, although it is more common to use the term to describe some subunit, or subsystem, of a building. In this chapter the issue of building systems is discussed with reference to the idea of a particular, distinctly definable means of construction. In that context, there are a small number of very ordinary systems that are repetitively used to create buildings.

6.1 HIERARCHIES

Systems with many parts often have the parts arranged in some ranked order or hierarchy (see Figure 6.1). For building construction a hierarchy may be defined in terms of the relative importance of elements in reference to some specific task or issue. Individual parts may be viewed as major or minor, primary or secondary, and so on. This may relate to considerations of the principal determinants of the system or to the relative significance of parts to design modifications for a specific purpose.

Modifications in the form of design variations may be more easily done when they deal only with secondary or minor elements of the basic system. Thus with the light wood frame, the studs are primary determinants of the system, while surfacing is secondary to the basic system. Many types of materials may be used for surfacing, effecting major changes in the appearance and various properties of the construction without essentially altering the basic nature of the structure.

The quality of adaptability just described for the light wood frame is one of the reasons it sustains its popularity; it can indeed adjust to almost endless variety in terms of application to different design styles and situations. There are a very few enduringly popular, basic construction systems that enjoy this quality.

6.2 Design Process and Tasks

A major early design task is the selection of the basic systems of construction for the building in general and for its major elements: roof, walls, floors, and so on. Options may be innumerable, or they may be shortened considerably by various design criteria, such as the construction budget, time available for design and/or construction, building code requirements, or strong client preferences. However determined, the firm choice of basic systems at an early design stage will greatly simplify the progress of the design work.

Figure 6.1   Articulated, hierarchical system: Steel columns and beams, light fabricated trusses, and a steel deck.

With basic systems defined, and their hierarchies understood, the forms of design modification that are most easily achieved can be explored. If the major design goals cannot be achieved within practical limits of the basic systems chosen, it is best to find this out very early. A shift in the basic systems has major effects on the progress of the work, and the earlier it happens, the better.

Although real design work seldom proceeds in so orderly a fashion, the following is a logical process for determination of the construction during the general development of a building design project:

1. Determine the basic systems. This means the selection of general forms and materials (masonry bearing walls with CMUs, for example).
2. Determine specific features of the basic elements (mortar type, basic unit type and dimensions, typical details, and other critical specifications for the CMU walls).
3. Consider effects on major building elements. This anticipates various problems of design integration. (How do unit dimensions affect window and door sizes, room dimensions, etc.?)
4. Develop all necessary major architectural details for use of the system. This tests the variability of the system and the ability to generate the range of situations that must be accommodated in the building, as well as the potential for alternatives.

6.3 MIXTURES

Buildings consist of large collections of individual parts and subsystems. There are very few manufacturers of whole buildings; in the main, each manufacturer produces only selected parts. The task of determining the collection of parts for a particular building is up to the building designer.

Practicality generally dictates that the parts of buildings are mostly obtained as manufactured products, selected from the catalogs or samples of the products as supplied by manufacturers or suppliers. In this situation, the building designer has the major task of comparative shopping and mixture making.

While a building represents a giant mixture, it is usually a mixture of subsystems and components, each of which may also be a mixture. For example, a designer may choose to use a light wood frame (2×4 wood studs, etc.) for a wall. This is a well-defined basic construction component and it can be "mixed" with various other components for an appropriate building assemblage. But the stud wall itself is a mixture, with studs produced by one maker, surfacing by another, connectors by a third, and so on. Each item can represent a design choice not necessarily implied by the general choice to use a wood stud wall system.

Selection between the available products for a single use (surfacing for a stud wall, for example) requires comparative value analyses that individual suppliers cannot be fully relied upon to assist with an unbiased view. The larger view of options, as well as real freedom of preference, is required. One type of wall surfacing may be the best in its class, with many superior qualities, but those qualities may be of minor concern for a particular application, and a lower-quality, less expensive product may be perfectly adequate for the limited task at hand.

Designers (shoppers, buyers) must look diligently for the best choice for each individual part for a building. But the whole building is more important than any individual part. The best roofing material, placed over a poor roof structure, will not produce a good roof.

6.4 APPROPRIATENESS OF SYSTEMS

The achievement of good construction begins with the choice of the appropriate materials, products, and systems for a particular use. Add in careful detailing, proper and thorough specifications, and competent construction work, and a well-built building just might result. But a well-built building in the wrong place or for the wrong use will still be inappropriate.

It is one thing to put plaster on a wall in the right way; it is another to select plaster as the appropriate surfacing for the wall for the situation at hand. Product information and standard specifications can deal with the correctness of the plaster itself, but the larger question of whether or not to use plaster is in the hands of the designer.

Manufacturers or industry associations sometimes provide guidance to the appropriate use of their products. However, they have in general a major interest in promoting the use of their products, so they do not much like pointing out all the situations in which those products are inappropriate.

Building codes provide some guidance for the selection of materials and systems for appropriate situations. This occurs mostly as a negative form of criteria—prohibiting certain uses in particular situations. These should be determined for any design work and used to narrow the choices very early on, but they typically leave room for many options.

Various unbiased reference sources exist for some assistance in making appropriate choices. However, individual buildings always have some specific qualified conditions that make simple evaluations difficult. Timeliness is also a major concern, as well as the many influences of regional situations.

In the end, the observations, experience, and judgment of the designer must prevail. This is what design is all about, and the task is humbling.

6.5 CHOICES: THE SELECTION PROCESS

Design of building construction is usually simpler and faster to achieve when choices can be made in large bites. Choosing preestablished (or essentially predesigned) systems is one means for doing this. If the choice is good in all regards, the size of the bite is irrelevant.

Preestablished systems offer the potential advantage of some demonstrated success. If the demonstrated successful cases have a reasonable match with the situation for a proposed design, all the parties concerned—designer, builder, owner, investors, insurers, and code-enforcing agencies—may take some comfort in an assurance of success. This is, of course, the actual case, making any real innovative design difficult to sell in many quarters. Designers should not be totally discouraged, but must understand the realities in achieving any significant changes in the "good old ways" of doing things.

Exciting, uplifting architecture can really be achieved with very ordinary means of construction, but can also—on occasion—come from real innovation or dramatic usage of the construction. In either case, the designer usually has a considerable understanding of, and appreciation for, the construction details and processes. If ordinary means are used, the designer understands their limits and potential, as well as what significant architectural design variations can be achieved within the basic systems. If ordinary means are rejected, the designer usually does so in full knowledge of what exactly is being discarded and what the needs are for any replacements. Otherwise, innovations will inevitably be naive and quite likely unsatisfactory in some performance aspects.

6.6 SYSTEM DEVELOPMENT CONCERNS

General planning of construction as well as choices for basic systems for an individual project must be made with the whole development of the project construction in mind. Certain portions of the work may be accomplished in a relatively separated activity for various reasons. One of these reasons is the separate functions represented by different design professionals.

Traditionally, architects were generally the master planners and coordinated the design work of most building design projects. This is still frequently the case, but the emergence of the separate activity of construction management and the role of developers has in many cases moved the command decision level to other people. In addition, the professional stature of many other members of the whole building design team has improved and strengthened. Interior designers, landscape architects, specification writers, environmental planners, and others have advanced the status of their professions, in many cases at the expense of some erosion of the leadership stature of architects.

In any event, the work of many strongly independent design groups must be coordinated for a successful project. In addition, the fractional division of the construction work is also a factor. This often results in the necessity to literally produce separate design documentation. For any project of significant size, there are the architectural drawings, the structural drawings, the electrical drawings, and so on. This presents a simple and necessary—but not easy—job of coordination of the multiple views of the design.

Continuing the problem is the fractional division of the building industry, with most manufacturers devoted to single materials or product lines. Thus design data, product detailed information, costs, availability, certifications, and other matters must be separately pursued. Industry organizations and standards are mostly quite narrowly defined and even regional in many cases.

The degree of fractionalization eventually overwhelms many attempts to coordinate the work, both for the design and for the construction. This is a major management problem and designers should be aware of it, but they should also realize that there are few *design* solutions for it. With regard to design, however, there are a few key relationships that should be given attention to be assured that an integrated design effort is achieved. These are treated in the following discussions.

Building and Site Planning

It goes without saying that the building and its site should be jointly developed as a continuous design concept. However, for large projects this involves the overlap of many separate professionals, and thus represents a major coordination effort. Simple vertical positioning of the building, for example, is a major decision with many potential consequences.

Preservation of existing site features and effective use of site materials can disappear as various design groups vie for supremacy in use of the site. Some real vigilance is often necessary for assurance of these efforts.

A key factor here is timing. Much design work and construction work is done in relation to the logical construction process: site clearing, excavations, and foundation construction early on, for example. In the now common use of fast tracking, design work and construction work are sequentially phased and overlapped to relate to a carefully timed construction sequence. This phasing puts items such as curtain wall detailing and landscape development work far down the line, with some risk of design judgments being made too late to have certain options if they were not anticipated very early.

As in many situations, the effectiveness of the designated master manager of the design work is critical in dealing with the problems of time sequence of the design and construction work.

Total Site Development

Many aspects of site development are controlled or even completely fixed by existing conditions, code requirements, or other factors. Exactly where a sewer line, fire hydrant, driveway curb cut, wheelchair entry ramp, or fire exit sign must be placed is not much subject to designers' judgments. Existing easements, property line setbacks, height of property line fences, and many other items may be fixed by ordinances, zoning restrictions, negotiated contractual terms, or other legal attachments or restrictions on the property title.

Experienced site designers and building designers know this well and very early in the design work make a major effort to establish all of these restrictive conditions.

Some items may be subject to negotiation and viewed as somewhat softer restrictions. Ones based on simple facts involving physical conditions are less flexible. The exact elevation of the sewer main that must be attached to, for example, is quite inflexible in most cases, if the line is already installed and serving many other customers.

Site and Foundation Development

A building site—in existing form or as viewed for development—has relationships dealing with both its finished surface and below-grade situations. Surface conditions must be related to site boundaries and to the general entry and exit from buildings. Below-grade items must be related to building foundations and to any special problems for the excavation work.

Choices regarding selection of the type of foundation, need for and methods used for bracing deep excavations, methods used for site dewatering (lowering of the site groundwater level), should be made in the widest context possible. This is often where the feasibility or simple accomplishment of preservation of site features can get lost in the shuffle.

6.7 INTEGRATION OF BUILDING SERVICES

Most buildings have electrical power, lighting, water supplies, waste piping, and HVAC systems. Multistory buildings have elevators, escalators, and vertical shafts for wiring, piping, ducts, chimneys, incinerators, mail chutes, and so on. Operation of the various services requires equipment that must be housed inside, on top of, or under the building.

Incorporation of service elements is a major chore in interior design, requiring considerable cooperation between the designers of the various separate systems. Overall coordination generally falls to the architectural designer and must be accomplished along with the design of the general interior construction. Some major considerations that must be made in this effort are the following.

Structural Interference

Structural interference relates to the problems of inserting all the nonstructural elements (including nonstructural construction) in a way that does not critically disturb the required functioning of the building structural system. Of typical concern are holes cut through walls or webs of beams, floor and roof openings that disturb orderly layouts of framing, and a need for horizontally continuous spaces that preclude locations for columns, bearing walls, shear walls, X-bracing, and so on.

The structure must function for safety of the building. If it cannot do so adequately and accommodate the other necessary building functions, something has to go. The structure is not inviolate, and some forms of compromise are to be expected. This is a two-way street, and a lot of judgment is required.

Fundamental Needs of Services

It helps if the designer understands some of the primary needs and limitations of the various building services. For example:

1. Waste drainage piping (from sinks, toilets, etc.) must facilitate gravity flow of liquids. This is especially a problem for long horizontal runs. A single pipe may theoretically fit inside some interstitial space, but its position must change if it has a long horizontal run.
2. Waste drains must be vented—generally through the roof and generally directly vertically as much as possible. This is a problem for the roof, as well as all the construction between the vented fixture and the roof.
3. Hot-water pipes and heating ducts get hot; chilled-water pipes and air-conditioning (cooled air) ducts get cold. Surrounding construction can be heated or cooled. Thermal expansion and contractions can create problems. Chilled objects can sweat and create moisture problems.
4. Electrical wiring, which must frequently be altered, and various other elements should be accessible for adjustments, repairs, alterations, and so on. Embedding elements in the permanent construction can create many problems.
5. Equipment with moving parts—notably motors, fans, pumps, and chillers—create both noise and direct physical vibrations that can be annoying. They should be isolated, possibly by good planning of their locations, but also by use of vibration isolators, sound-insulated partitions, or other special construction.

As noted previously, compromise is the name of the game. Ideally, all building systems should operate or exist with optimal conditions, but economic feasibility, technical limitations, and some priority of design values and goals must be recognized.

6.8 FACILITATION OF MODIFICATION

A peculiar requirement of modern buildings is their general need to facilitate modification. The dynamic nature of our society makes this imperative. People relocate regularly; businesses reorganize; the urban fabric continuously changes—growing, decaying, regrouping; and rapid changes in technology bring many sudden demands for facilitation of new opportunities.

Playing against this rapidly shifting scene is the great expense of building costs, which makes the writing off of initial construction over a long period mandatory for most investors. The net effect is that buildings must be built for permanence but somehow be easily, feasibly, and preferably rapidly modifiable.

Ease of modification is a design goal and involves both basic planning and the final decisions for choices of materials and details of the construction. It also needs to permeate the whole design effort, including that of the structure and all the building services where possible. This is relatively easy for some things, such as wiring, ceilings, and plumbing fixtures, but not so easy for sewers, foundations, and shear walls.

Designing for special requirements, such as ease of modification, may require extra effort and extra cost of construction. In some ways, however, it merely means picking between competing alternatives with a view to those that facilitate ease of change. Bolted joints can be undone, for example, while glued or welded ones cannot.

7

Case Studies

The purpose of this chapter is to present a number of examples of building types, construction usages, and site situations and to illustrate the development of the general construction for each example.

The presentations here are meant to complement the materials in preceding chapters. Critical isolated issues are dealt with more extensively in those chapters, and that material should be used as a resource for the more detailed information on individual items shown in these examples.

However, designers must ordinarily face the problem of the whole site and the building/site relationship, even though the actual design may be worked out by concentrating on one problem at a time. Here, the presentations begin with consideration of the building's general form and construction system, followed by presentations of some of the primary increments of the construction with all of the significant parts displayed.

The building designs presented here are not meant as illustrations of superior architecture. The interest here is directed mostly to setting up reasonably realistic situations that cover a range of circumstances in regard to design of construction. It is not the intention to present boring, ugly buildings, but the use of a limited number of examples to display a broad range of construction use is the dominant concern.

An attempt has been made to show construction usage and the form of details that are reasonably correct. This is, however, a matter for much individual judgment and is often tempered by time and location. There are not many situations in which only one way is correct and all others are unequivocally wrong.

Reasonably correct alternatives are often possible, and the particular circumstances for an individual building can make a specific choice better on different occasions. Differences in climate, building codes, local markets, building experience, or community values can produce a way of doing things that is preferred. Over time, changes in any of those areas of influence will produce variety and hopefully some evolution toward a better way of doing things.

The concentration in this book is on sites and the below-grade portions of buildings. General form and construction details of the buildings are presented here only to the extent necessary to understand the development of the site as a setting for the building. The general development of the construction for these examples is considered in the two books which are the first volumes of this series—*Building Construction: Enclosure Systems* and *Building Construction: Interior Systems*. Some of the details of the buildings that are developed in those books are shown here so that the reader can more fully understand the context of the interface between the building and the site.

Finally, although incorporation of building services is a major concern for the development of the building/site interface, the full development of these systems cannot be shown in detail here. However, some critical issues that significantly affect individual elements of the building construction or present problems for site development are discussed in the examples.

7.1 BUILDING 1

Single-Family Residence
Light Wood Frame
Sloping Site

This is a building of modest proportions; not impressive to architectural critics, but a castle by world housing standards. As a type of building—and in this general size and general character—it is probably the most widely constructed building in the United States. Cape Cod cottages used this basic structure hundreds of years ago and are still imitated as a style in extensive housing developments. The form of house shown here, in Figure 7.1, is that of a typical suburban split-level, but many details are really reminiscent of the basic Cape Cod cottage.

Figure 7.2 shows the plot plan with the building indicated only by its overall "footprint" on the site. The building/site sections show the placement of the building vertically in relation to the sloping site. On both the plan and section drawings the finished construction and finished site grade are shown in hard-line profile, while the original site form is shown with dashed lines.

It may be observed in the sections in Figure 7.2 that the building base steps down the site somewhat, but that some cutting into the site at the back of the building and some building up with fill at the front are required. Some construction details required to achieve this are shown in Figures 7.3 and 7.4. If possible, it is desirable to place the building so that the amount of excavated or cut materials is approximately equal to the fill required for surface buildup within the site boundaries. This will eliminate the need for either removal and disposal or importing of major volumes of soil.

Note on the site plan that the regrading work fades out at the site edges, so that changes occur only within the site boundaries. This is not always possible with tight site spaces, and it may require the building of retaining walls or other structures at some edges. The general problem to be solved is that of meeting other properties at the edges—including the existing streets, curbs, and any sidewalks.

Not a small problem for any site is that of recontouring in a way that does not unintentionally channel precipitation runoff onto the neighbors or with concentrated velocity onto the street. Covering the site with a large building and a lot of paving may well result in more total runoff than occurred with the barren site, further aggravating the problem. Once a preliminary site plan is developed, a careful study of the site drainage should be done.

The site drainage problem can also be aggravated by the installation of an extensive irrigation system, producing in some cases more total runoff than that occurring with precipitation. Avoiding this requires careful design of the irrigation system, a study of site drainage from this source, and a lot of thought about the landscape design. If this is a major potential problem (which hillside houses typically present), the selection of water-conserving plantings, use of trickle systems for watering, and other methods should be used.

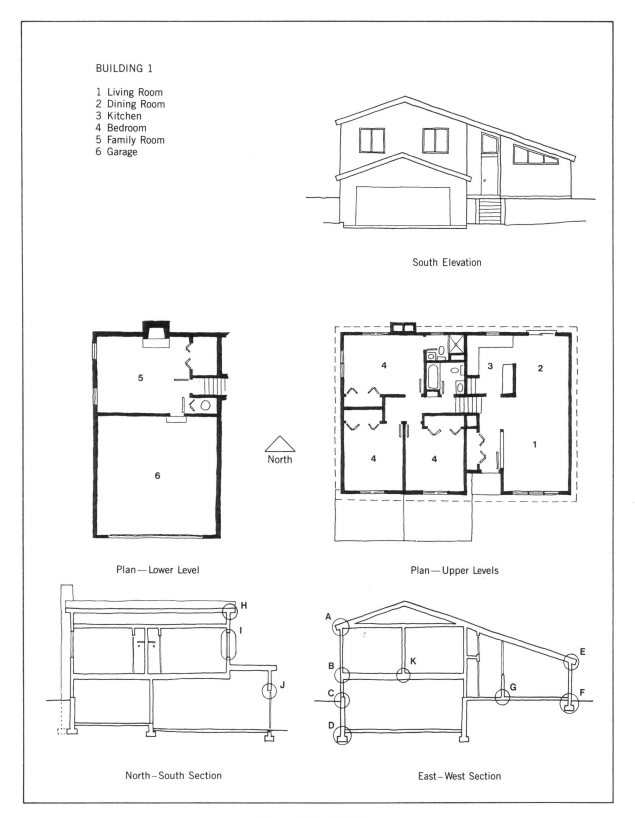

Figure 7.1   Building 1.

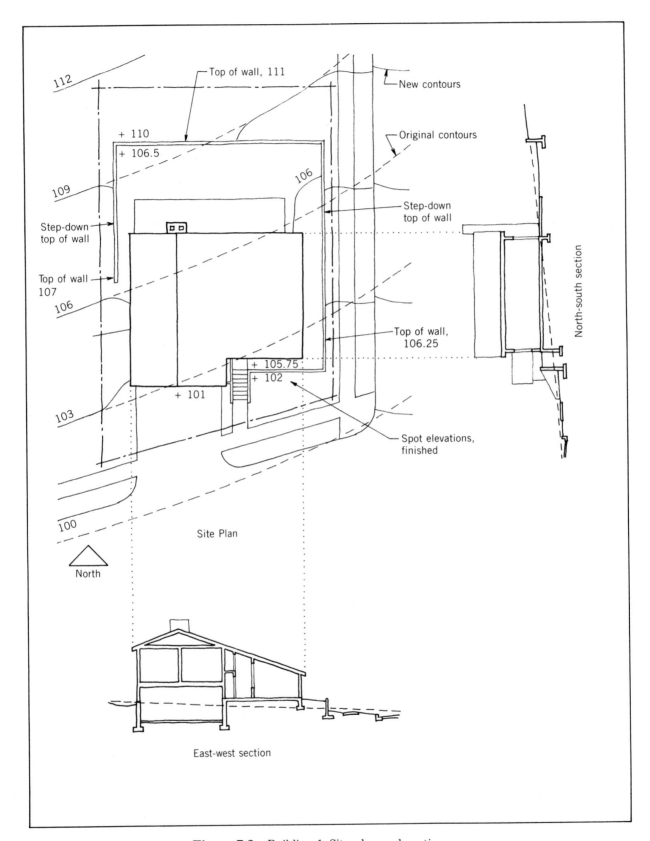

Figure 7.2   Building 1: Site plan and sections.

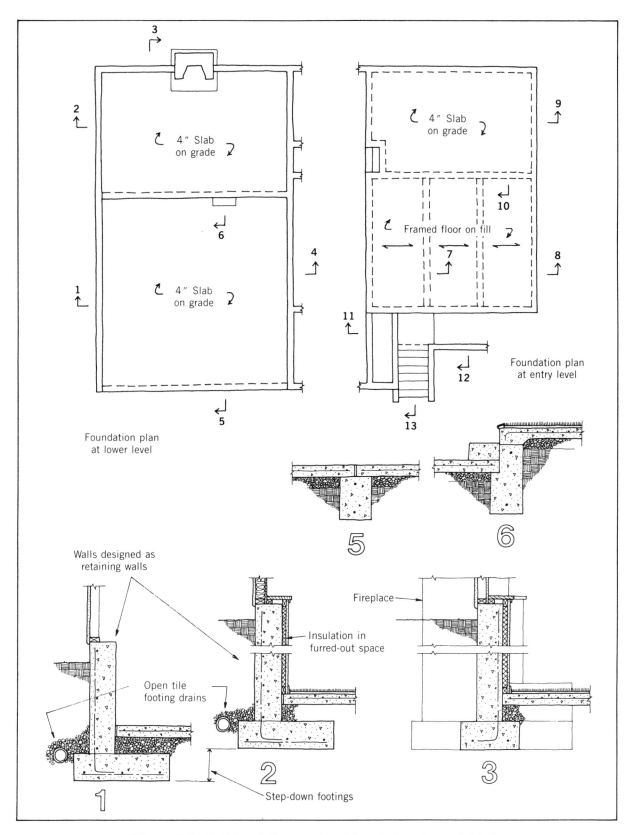

Figure 7.3 Building 1: Basement and foundation plan and details.

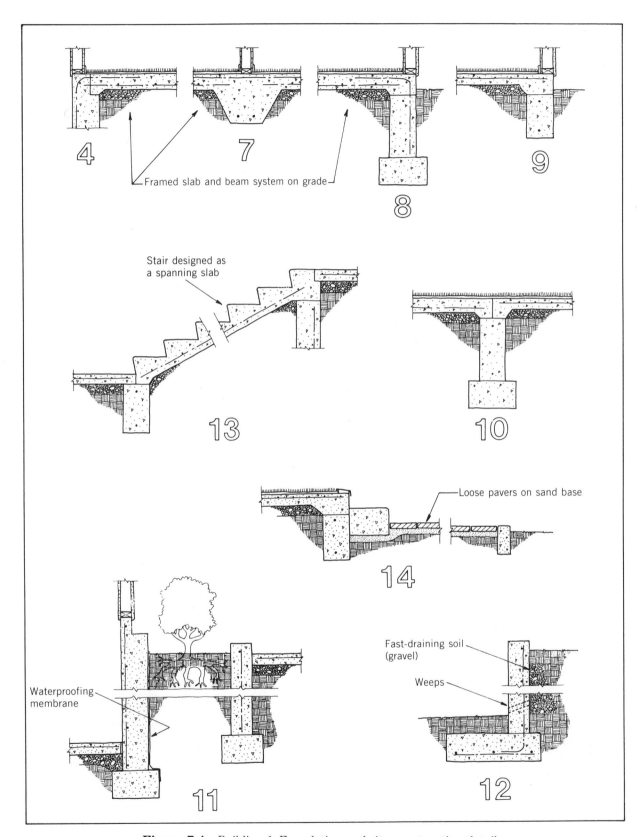

Figure 7.4   Building 1: Foundation and site construction details.

The plan of the building base is shown in Figure 7.3. This indicates the bounding foundation walls in the lower part of the house, the concrete floor paving slab in the entry-level part, and the plans (in dashed lines) of the foundations. Inspection of this drawing together with the site plan and sections will reveal the various conditions with regard to the positions of construction elements relative to both the original site surface and the finished grades.

The walls at the sides of the garage and on the north and west sides of the lower level must retain some soil pressures against their outer sides. This requirement presents a special problem because the concrete portions of these walls are not supported laterally at their tops in most cases. This makes it necessary to design these walls as cantilever retaining walls or as horizontal-spanning ones in a lateral direction. The details in Figure 7.3 show the use of a form of retaining wall construction, although the bracing by the floor slabs is also critical to the wall stability.

At the front end of the house, the floor in the living room and entry areas, the front walk, and the exterior stair are placed on considerable fill. The plans and details show the use of spanning concrete construction at this location. Note that all foundations are taken down to a level that is slightly below the original site surface.

At the rear of the house the opposite problem occurs and it is necessary to cut into the original grade to provide the flat patio area. This also requires the construction of some form of retaining structure on three sides of the patio. As the details in Figure 7.4 indicate, a short concrete cantilever wall is placed across the site at the point of the deepest cut. This wall is returned a short distance at its ends, but the remainder of the retaining is done with a loose-laid stone wall, sloped into the cut. Various other options for this type of structure are discussed in Section 4.6.

7.2 BUILDING 2

One-Story Commercial Light Wood Frame Flat Site

This is a small street-front building with parking in front and access to the rear through an alley as well as around one side. Assuming that the site is approximately at its original level, it should be possible to use simple paving on grade for the drives, parking, walks, and building floor. Building footings must be placed at a depth relating to local frost depths—usually a minimum distance as specified by local building codes.

Figure 7.5 shows the general form of the building. A canopy cantilevered from the front wall partly overhangs the walk at the front of the building. This walk is slightly buffered from the parking in front by a row of low planters, as shown in the plan and sections in Figure 7.6. Other site features include the low wall at the edge of the front sidewalk and the tall, property line wall at the west edge of the site. Details of this construction are shown in Figure 7.8.

The building section in Figure 7.5 shows that the roof pitches to the rear of the building. Scuppers are located at the rear, through the parapet wall, and feeding into vertical leaders on the rear wall. These in turn feed into vertical tile drains at the ground and into a sewer along the back of the building. This sewer may join with the sanitary sewer for the building or feed into a separate storm sewer.

Other site surface runoff is assumed to be into the street and alley, if permitted by

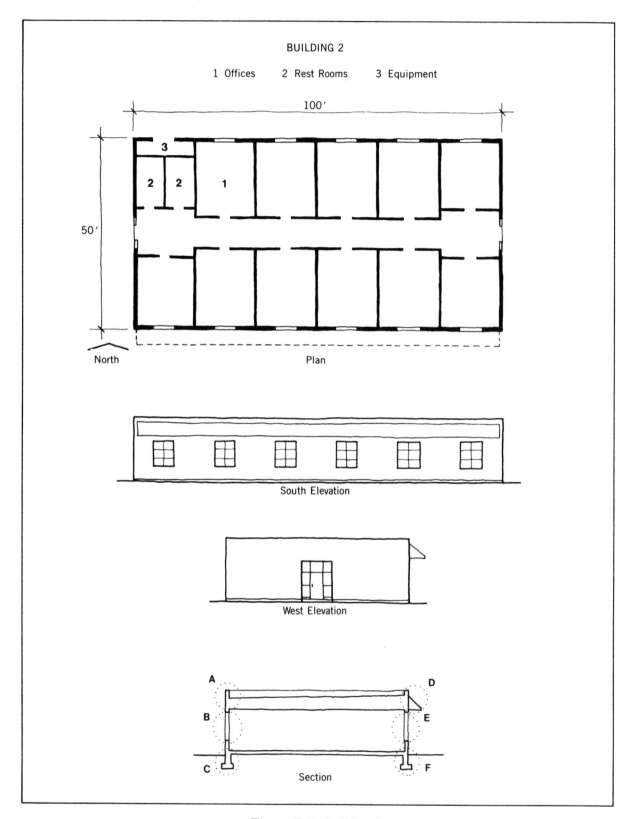

Figure 7.5 Building 2.

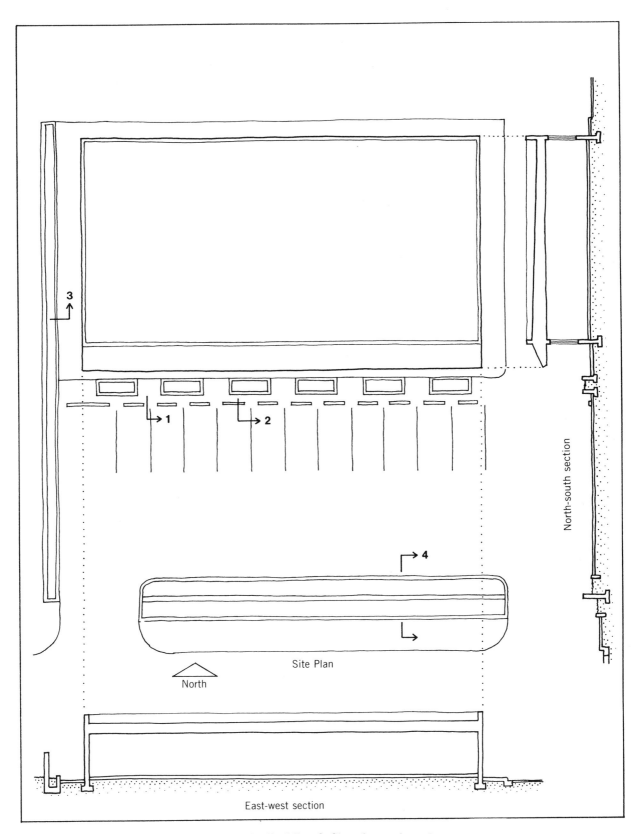

North-south section

North

Site Plan

East-west section

Figure 7.6 Building 2: Site plan and sections.

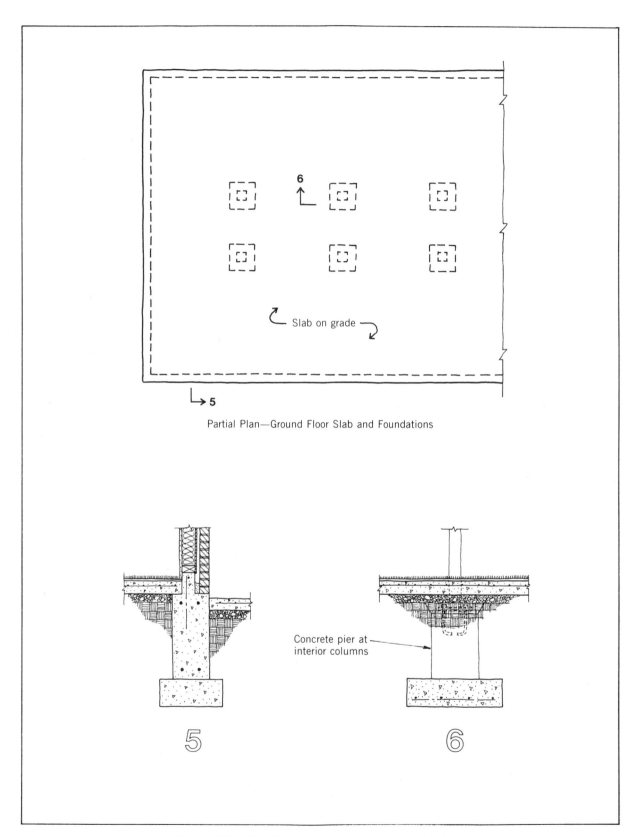

Partial Plan—Ground Floor Slab and Foundations

Concrete pier at
interior columns

Figure 7.7   Building 2: Foundation plan and details.

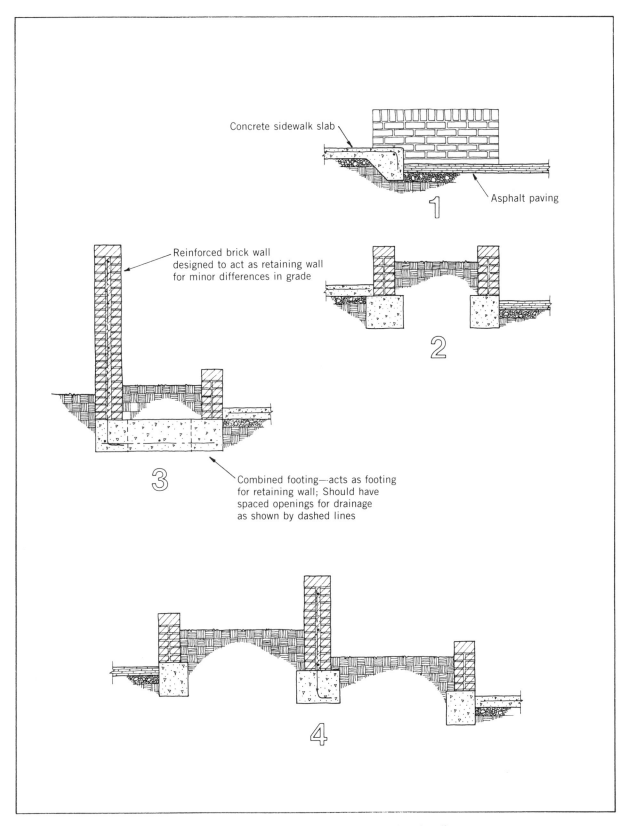

Figure 7.8 Building 2: Site construction details.

local agencies. Locations of drains in the alley and street would form some references for development of the site contours to achieve proper drainage.

A portion of the building foundation plan is shown in Figure 7.7, along with some details of the grade beam foundation. The foundation grade beam/wall is used to distribute the wall loads to a continuous-width footing, even though some loads are widely spaced along the front, due to the large openings.

7.3 BUILDING 3

One-Story Industrial
Concrete/Masonry Walls, Frame Roof
Large Flat Site

This is a multiple-bayed building, indefinitely extendable in two directions, producing a large covered floor space for an industrial plant, warehouse, or other such usage (see Figure 7.9). Depending on requirements for clear spans, wall heights, height beneath the roof structure, and total floor area, various structural elements may satisfy code requirements, user needs, and the general desire for very economical construction.

As shown here, the roof has a slow-draining, almost flat profile. Drainage of such a roof is a major problem as areas near the center of the large building are some distance from a building edge. Edge draining, as shown for building 2, is generally not feasible here, as the total amount of the pitch from the center to an edge would be prohibitive. Some edge draining may be provided, but some roof drains of the type used for interior drainage will be required, most likely discharging into interior vertical leaders located at columns and in turn feeding into an underfloor piped sewer system.

The general form of the building is shown in Figure 7.9. Two possible schemes for the construction are shown. One employs precast concrete wall panels and a steel roof system; the other has walls of reinforced CMUs and a wood roof system. Both schemes use simple bearing footings and a floor slab on grade.

The site plan in Figure 7.10 indicates a flat site and shows some of the connections to services. Water, gas, and sewer lines must be underground, involving some careful considerations in site planning. Sewers are a particularly difficult problem as they must drain by gravity flow. If the elevation of the sewer main to which attachment is made is not substantially lower than the building, and the distance is great between the building and the main, special provisions must be made. To illustrate the problem, some details are given in Figure 7.12 for a system used to raise the collected building waste water to a level permitting it to drain to the sewer main.

Electrical power and telephone lines could be delivered above ground, but are also shown here in an underground delivery system. Various forms of access to these services may be required, involving the installation of underground vaults, tunnels, and so on. The locations as well as details for the construction must be coordinated with the various agencies or companies providing services. For a large project, these may constitute major site elements which should be planned for as well as possible.

The building/site section in Figure 7.10 also shows a truck dock at one side of the building, a common feature permitting easier loading and unloading of large trucks. A depressed drive area is shown on the plan and section to permit use of this floor-level dock. The general form of such a dock is shown in the detail section in Figure 7.10.

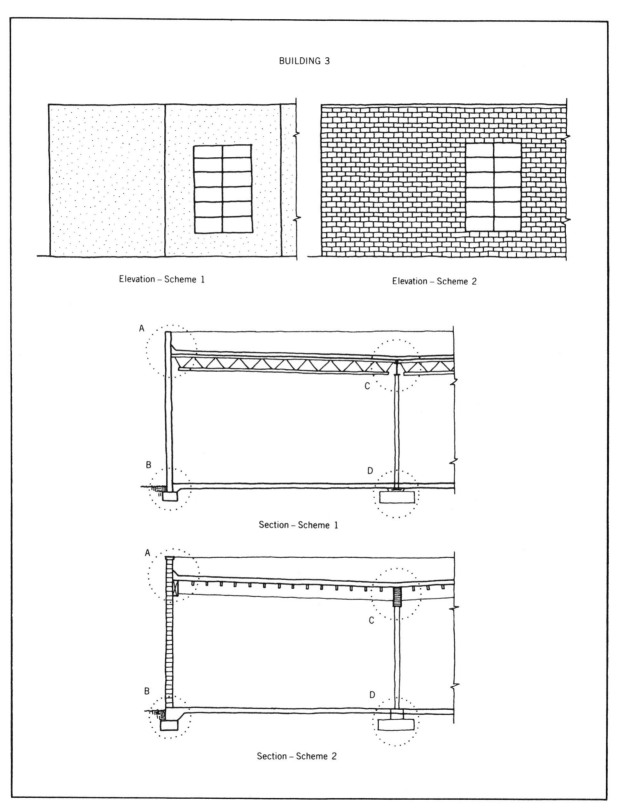

Figure 7.9 Building 3.

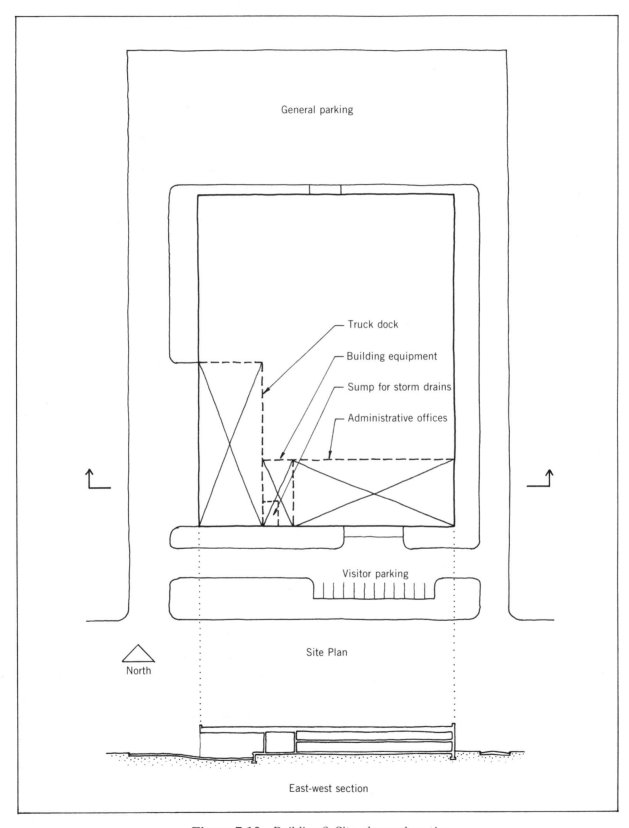

General parking

Truck dock

Building equipment

Sump for storm drains

Administrative offices

North

Visitor parking

Site Plan

East-west section

Figure 7.10   Building 3: Site plan and section.

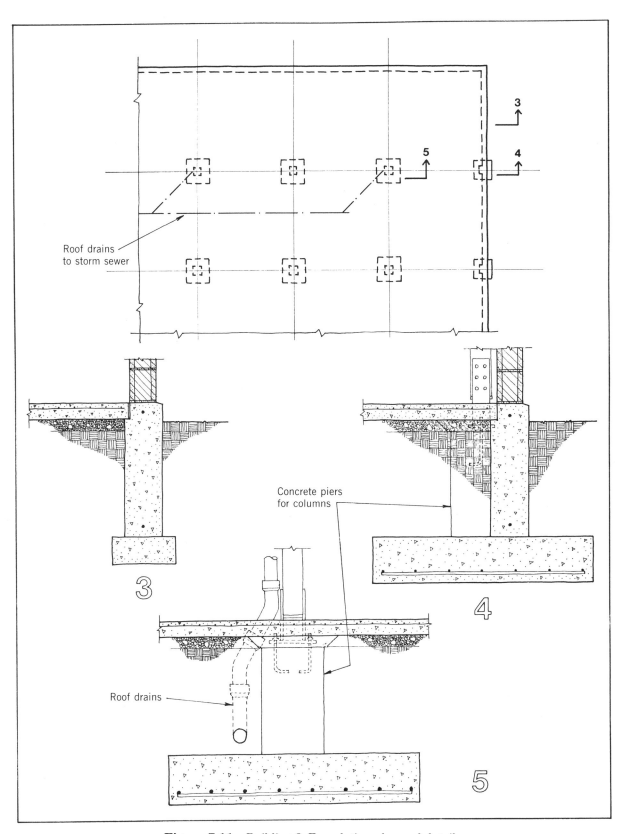

Figure 7.11   Building 3: Foundation plan and details.

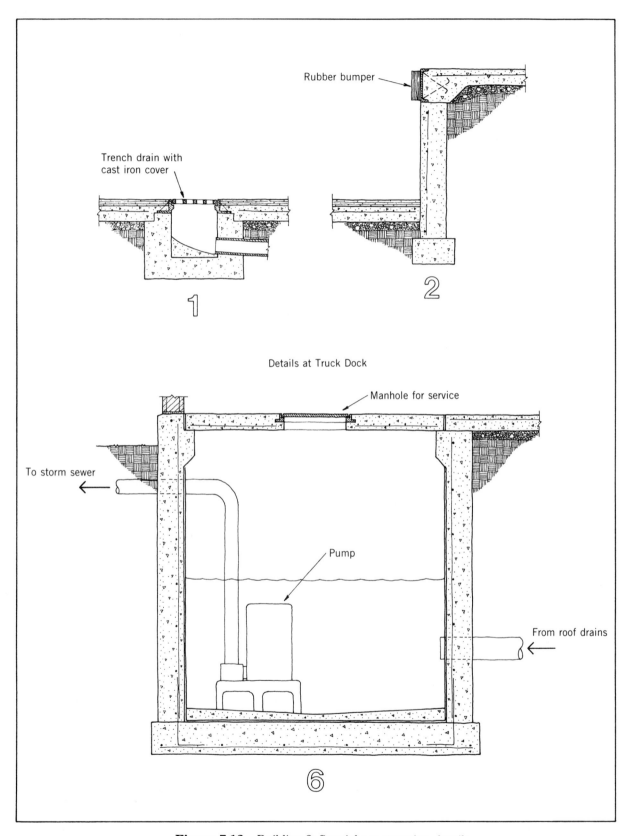

Figure 7.12   Building 3: Special construction details.

A partial foundation plan is shown in Figure 7.11, together with some details of the foundations. A grade beam/wall foundation with a continuous footing—similar to that for building 2—is used at the walls. These are quite heavy walls, so the footing is likely to be somewhat wider than that for building 2.

The footing for the interior column is shown with a concrete pedestal that is used to raise the column base to the floor level—a general requirement for wood or steel columns. Also shown with this detail is one of the interior roof drains that feeds into the underfloor sewer system.

More details of the underfloor drain system are shown in Figure 7.12. With the large horizontal dimensions of the building, it is likely that the required gravity drain slope of this sewer will place its lowest point considerably below grade level before it exits the building. Assuming that it is also some distance to the sewer main off the property, this can create a problem. Shown in Figure 7.12 is the use of a sump pit and pump, used to raise the sewer wastewater to a level closer to the ground surface before it leaves the building.

7.4 BUILDING 4

Two-Story Motel
Masonry and Precast Concrete
Sloping Site

As shown in Figure 7.13, this building has a common form of plan for motels, with rooms off a central corridor. The same basic structure might also be used for a dormitory or small apartment building.

The basic structure consists of structural masonry walls and precast concrete roof and floor units. Although this is a widely used construction, there are many possibilities for variations, both in the general structure and in finish materials and details.

The site plan and sections in Figure 7.14 show the sloping site and the various features used to develop the parking areas, drives, and entries to the motel. Note the adjustments in site planning used to accommodate several existing trees. Protection of one of these trees requires the construction of a well-like retaining structure around it, details for which are shown in Figure 7.16. Details for other site construction are also shown in Figures 7.16 and 7.17.

A partial foundation plan for the building is shown in Figure 7.15. The ground floor of the building is a concrete slab on grade and there is no basement, so the structure makes a minimal penetration of the site for general construction. However, the use of elevators requires some construction to a depth necessary to achieve the elevator pit, details for which are shown in the figure. It is quite possible that these would be hydraulic elevators, requiring typically some major ground penetration for the hydraulic piston mechanism.

Also shown in Figure 7.15 are some details for the insulation of the building edge at ground level, as would be desirable for cold climates to prevent cold floors.

Figures 7.16 and 7.17 show details for several site elements that are identified on the site plan in Figure 7.14. Shown on the plans, but not here, are several other elements, including walks, stairs, and retaining structures. Examples of the latter are shown for several of the other building examples.

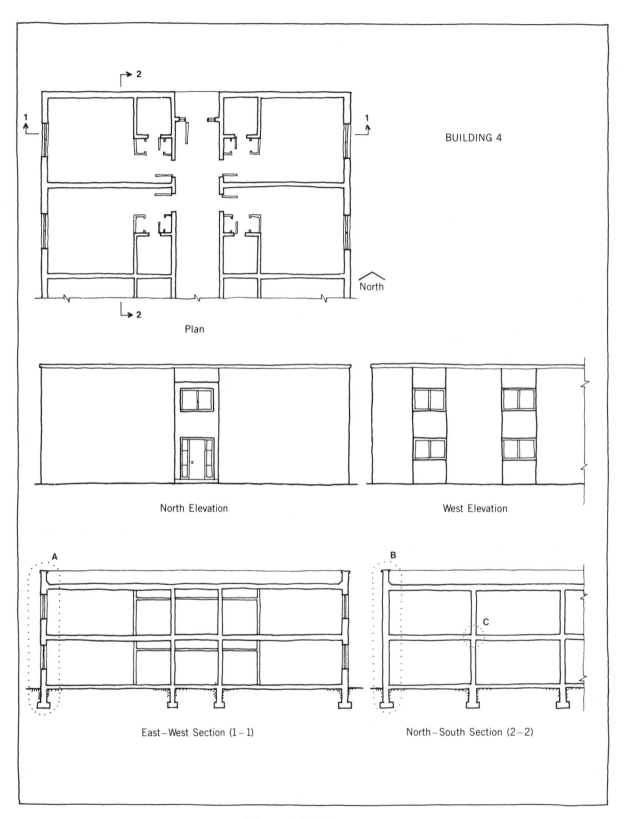

Figure 7.13 Building 4.

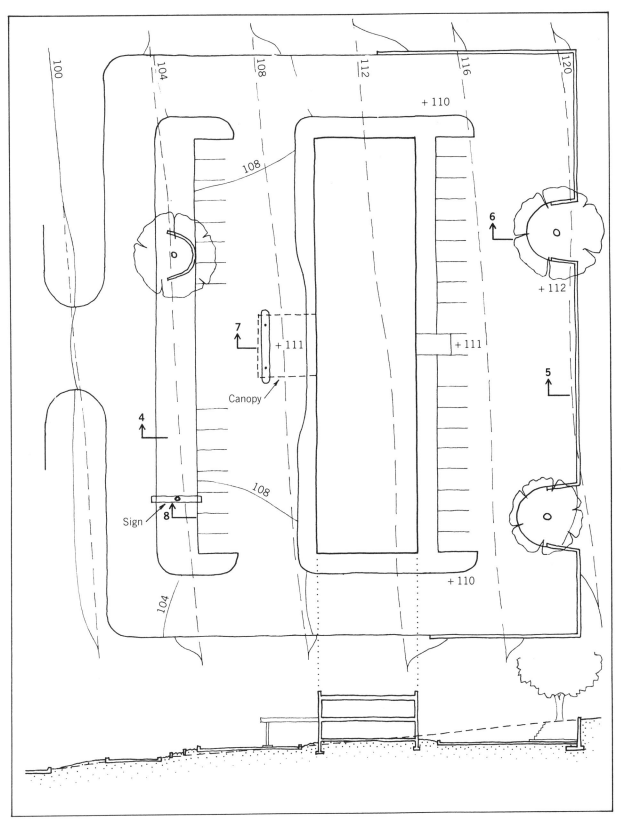

Figure 7.14   Building 4: Site plan and section.

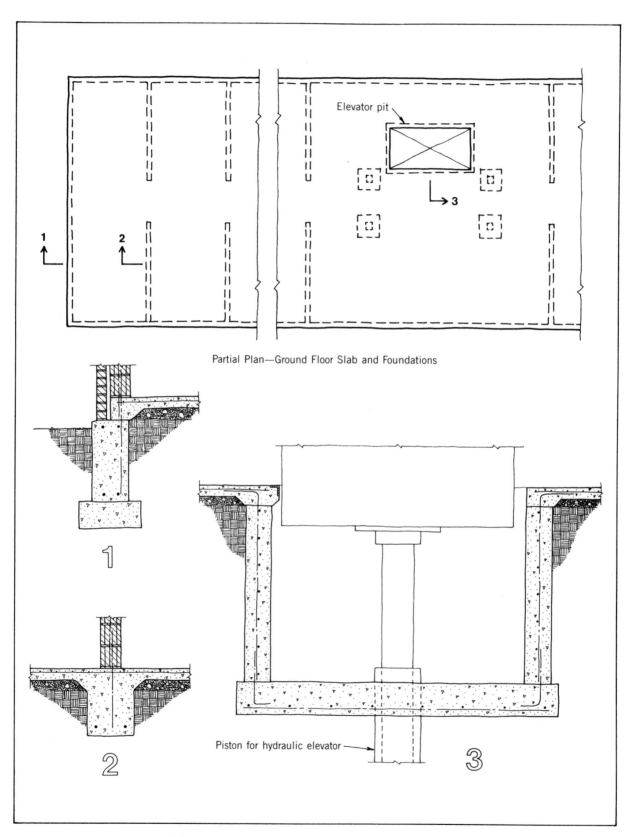

Partial Plan—Ground Floor Slab and Foundations

Elevator pit

Piston for hydraulic elevator

Figure 7.15   Building 4: Foundation plan and details.

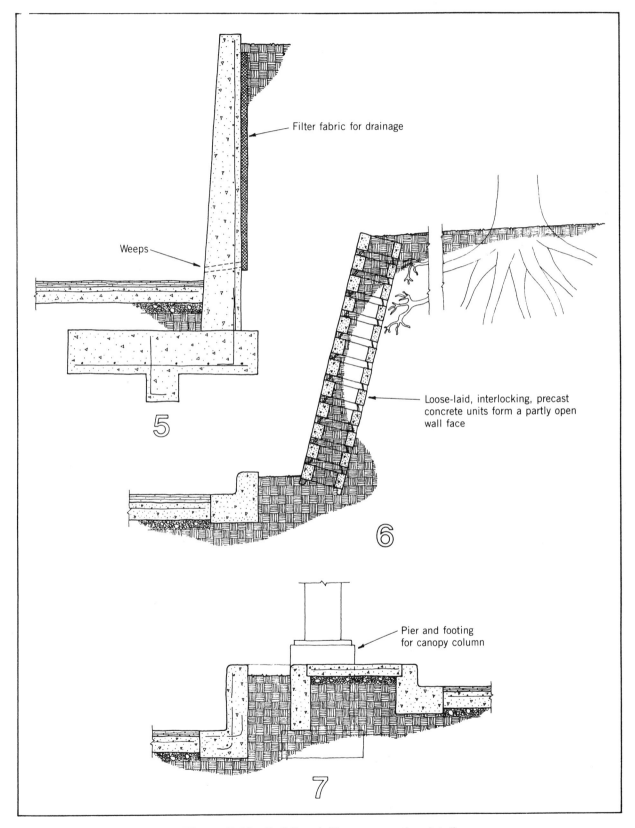

Figure 7.16 Building 4: Site construction details.

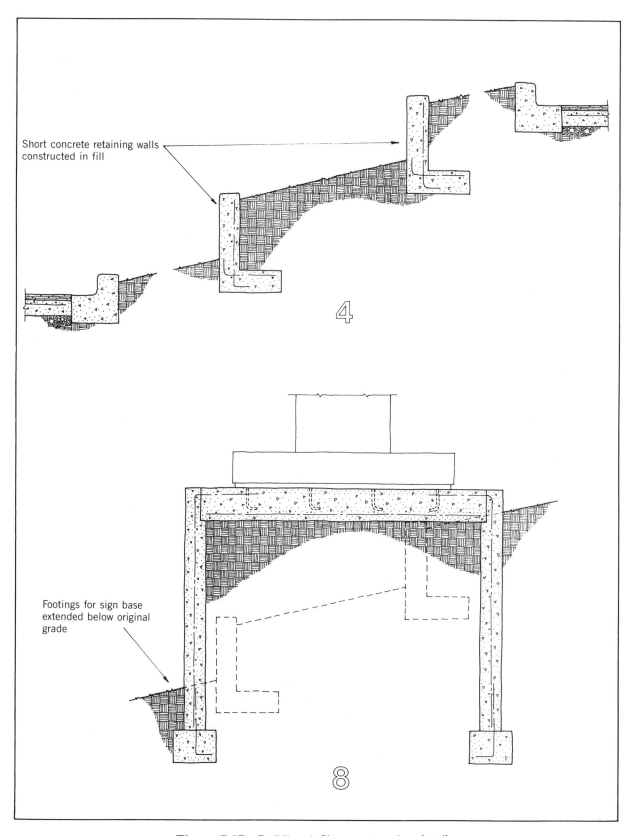

Figure 7.17   Building 4: Site construction details.

7.5 BUILDING 5

Church Auditorium/Chapel
Exposed Timber Structure
Flat Site
Part of a Building Complex

This is the principal building of a small complex that includes a residence, a classroom/office building, and an extensive parking facility. The buildings are discrete structures, but are linked with outdoor spaces, structures, and landscaping.

The main church building itself is a single-space, medium-span building with essentially no interior spatial division—an enclosing shell for housing of singular events. These events may include a range of types, from very formal liturgical services, funerals, and weddings, to general meetings and social events for the church members.

To utilize the site space efficiently, there will be some pressure to economize on outdoor space not used directly for parking, buildings, and necessary walks and driveways. Still, some setting should be achieved for the church and some space available besides the parking lot for outdoor assembly and casual meeting before and after events. It may also be important to have the entry into the church be one that occurs as a procession through some transitory spaces. It is assumed here that the spaces between and around the buildings will be developed with these purposes in mind.

The character of exterior construction should relate to that of the church building. As shown in Figures 7.18 and 7.19, the church is constructed with an exposed timber frame, a tiled roof, and stucco walls. This rough, rich texture and the natural wood finish should be a reference for selection of exterior finishes.

The site plan in Figure 7.20 shows the general arrangement of the complex. The church is featured centrally on the site and is further emphasized by raising it slightly on a built-up fill, a few feet above the rest of the flat site. Walks, landscaping, and an entry court also serve to create a transitional entry and visual spaciousness for the church building. The site section in Figure 7.20 shows both the horizontal and vertical form of this spatial arrangement.

Figure 7.21 shows a partial foundation plan for the church building. The concrete floor slab and exterior walls are supported by a grade beam and wall footing that fits between the piers for the frame bents. The details for the piers show the use of one method for developing the resistance to outward thrust at the base of the bents. In this case, the heavy floor slab is used to anchor reinforcing extended from the piers. Another solution is to provide a continuous tie across the structure, from pier to pier on opposite sides. The slab drag anchors shown are generally simpler and considered adequate for such structures when the span is modest, as it is here.

Since the floor of the church is considerably above the level of the original site, there is some judgment to be made about its structure. Use of a simple slab on grade would require very careful placing and compaction of the fill used to achieve the slab base, all the way up from original cleared grade. If this is not considered feasible, a framed floor on grade, such as that used in the front area of building 1, must be used to assure a stable floor structure.

Figure 7.21 also shows some details for the construction of the raised altar area at the front of the church, using a concrete structure and stone or tile paving. This form could also be achieved by placing a flat floor slab and building a wood platform on top

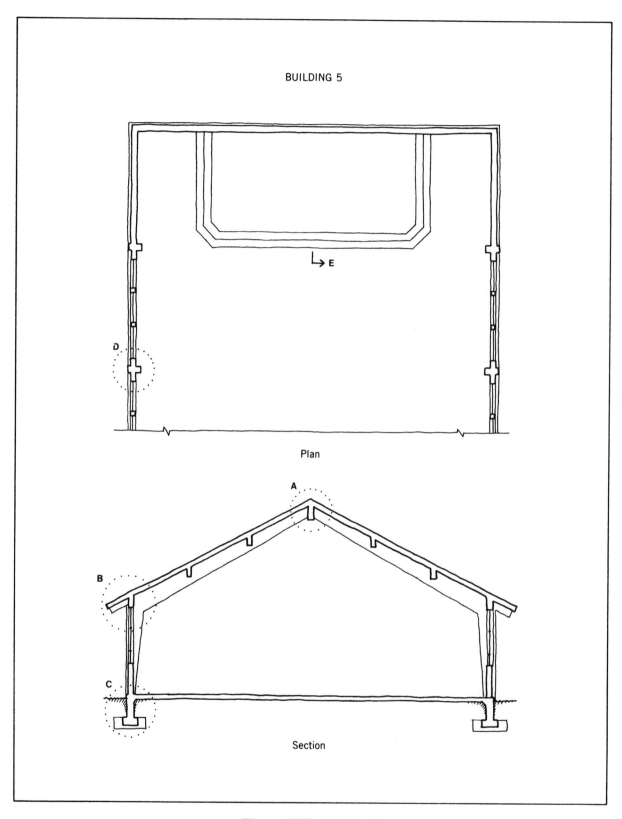

Figure 7.18   Building 5.

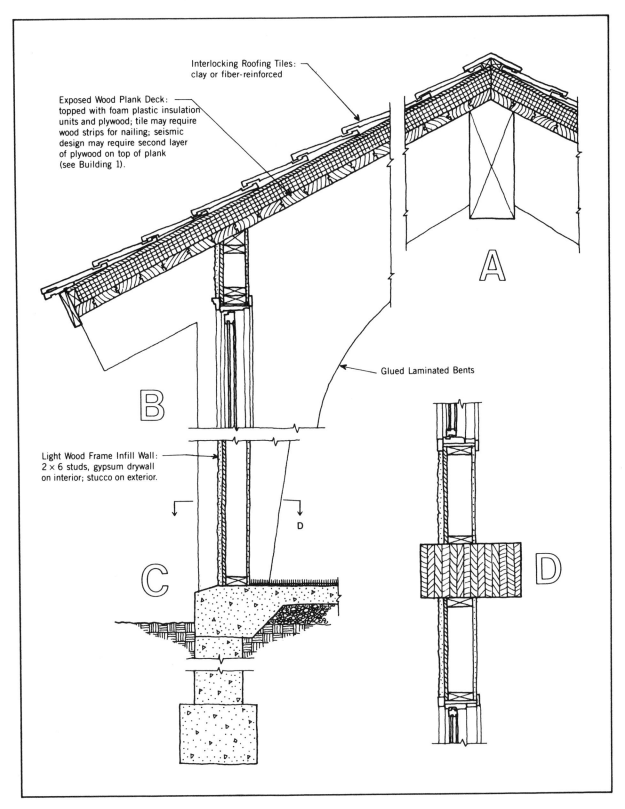

Interlocking Roofing Tiles: clay or fiber-reinforced

Exposed Wood Plank Deck: topped with foam plastic insulation units and plywood; tile may require wood strips for nailing; seismic design may require second layer of plywood on top of plank (see Building 1).

Glued Laminated Bents

Light Wood Frame Infill Wall: 2 × 6 studs, gypsum drywall on interior; stucco on exterior.

A

B

C

D

Figure 7.19 Building 5: Construction details.

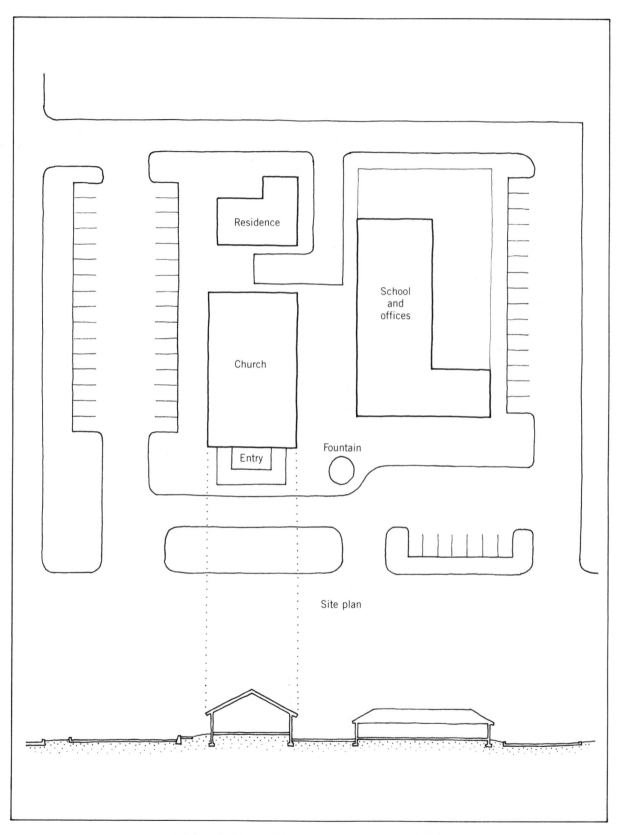

Residence

School
and
offices

Church

Entry   Fountain

Site plan

Figure 7.20   Building 5: Site plan and section.

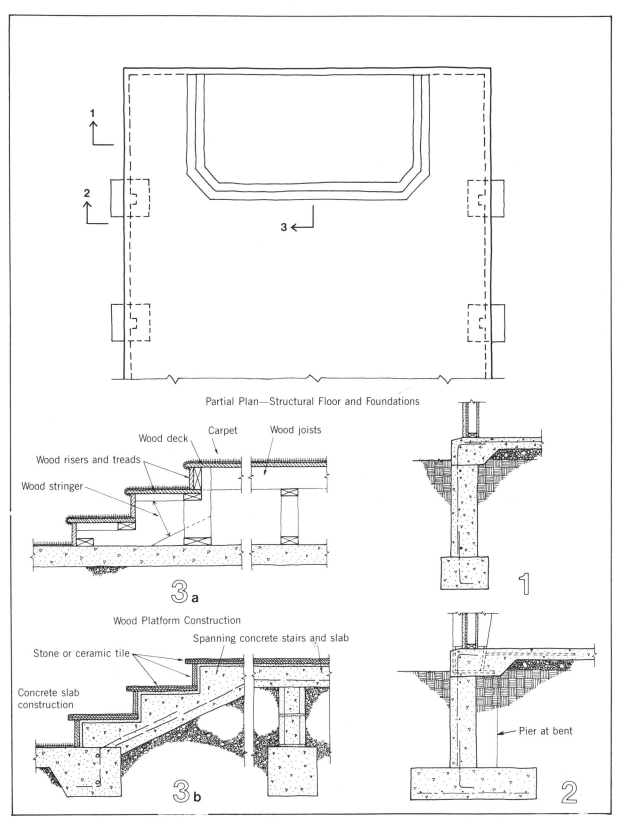

Figure 7.21   Building 5: Structural floor and foundation plan and details.

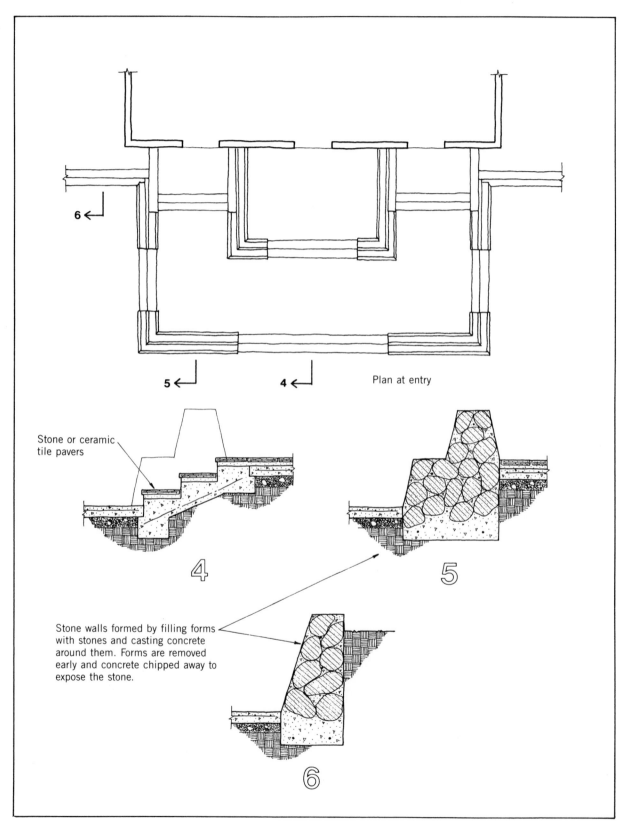

Plan at entry

Stone or ceramic
tile pavers

4

5

Stone walls formed by filling forms
with stones and casting concrete
around them. Forms are removed
early and concrete chipped away to
expose the stone.

6

Figure 7.22   Building 5: Entry and site construction details.

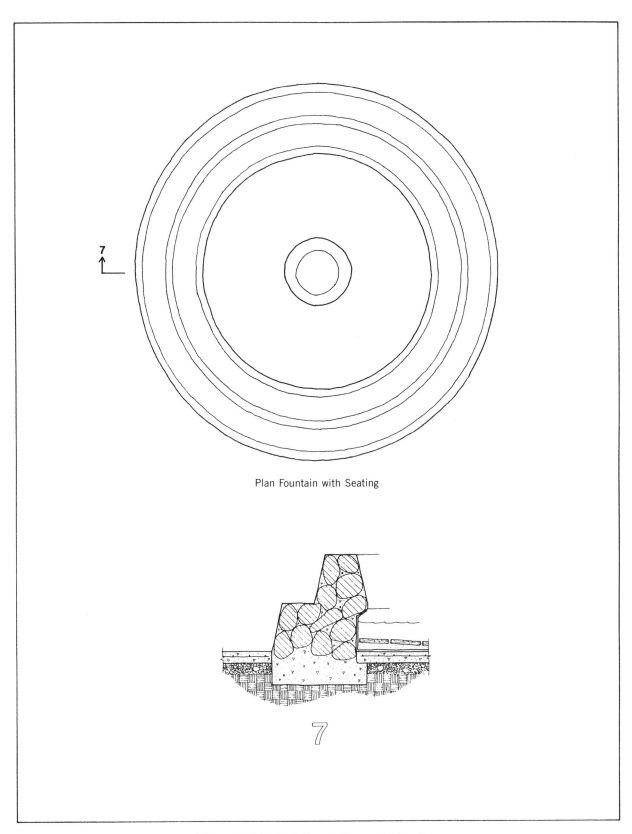

Plan Fountain with Seating

7

Figure 7.23 Building 5: Fountain details.

of it—probably a less expensive construction, but not as good a base for the tile or stone.

Figures 7.22 and 7.23 show the construction of some of the site elements outside the church. Carrying out the general nature of the rough, natural materials of the church, these are developed with stone, tile, and timber as much as possible. Low walls are developed with large stones in a strongly tapered, truncated pyramidal form, reminiscent of stone piles achieved without mortar or concrete by early builders. Even though modern concrete is used here to tie the construction together, the stone work should still be carefully developed so that the rock piles are as stable as possible without the concrete.

In keeping with other materials, the exposed concrete work here could be done with a concrete with a tan tinting admixture and possibly some surfaces brushed or air blasted to expose the aggregate.

Some walk and terrace surfaces are shown with paving of stone or tile. In some areas stone paving could be done with loose-laid pavers over a compacted granular base. Where the base soil is considerably built up, this form of paving can probably suffice without concern for surface drainage—relying on a very fast-draining subbase for the pavers. If necessary, an open tile drain system could be placed beneath the paving, draining out to the lower parking areas.

In keeping with the details of the church roof structure, wood is used for the benches around the fountain, and some use of treated timber edging might be used for some planting areas. The fountain and pool are somewhat questionable for a building that is used only periodically, but would be more feasible if general continuous use is made of the church complex.

A general concern for such a complex is that of continuous maintenance, which may be mostly by volunteers from the church membership. These workers may be motivated and industrious, but they are not likely to be professional landscape maintenance people. Thus, the choice of plantings and general landscape elements should be made with simple maintenance as a major concern.

7.6 BUILDING 6

Three-Story Office Building
Steel Frame, Metal Curtain Wall
Large Sloping Site

This is a modest-size building, generally qualified as being low rise (see Figure 7.24). In this category there is a considerable range of choice for the construction, although in a particular place, at a particular time, a few popular forms of construction tend to dominate the field. Currently available products, market competition, current code requirements, popular architectural styles, and the general preferences of builders, craft people, developers, and designers usually combine to favor a few basic forms of general construction.

Shown here is a steel frame structure with W sections used for columns, girders, and beams, and a formed sheet steel deck for the floors and roof. The exterior wall system is a curtain wall developed as an infill steel stud system. Windows consist of strips between columns, framed into rough openings in the stud wall structure.

Figure 7.25 shows a partial plan of the ground floor at the location of the building core—a clustered group of facilities including stairs, elevators, duct shafts, and

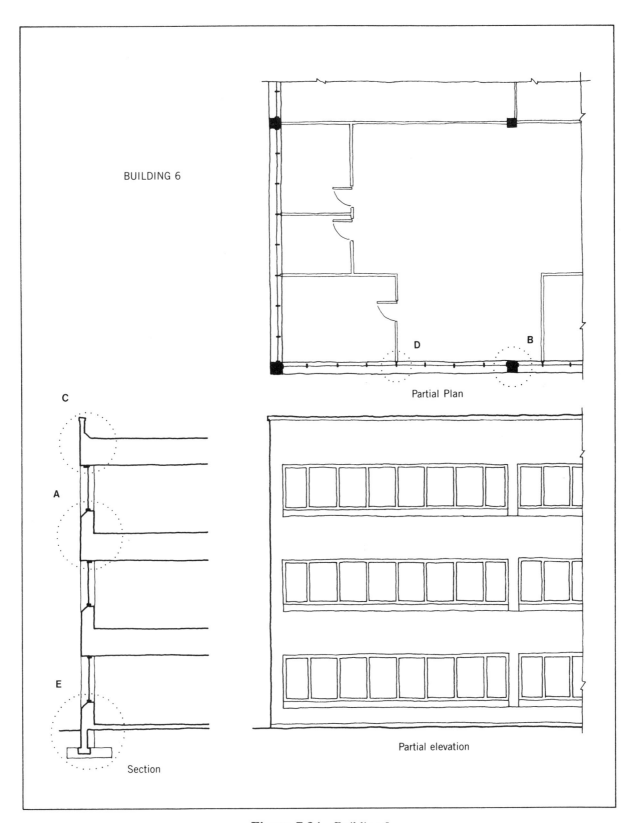

BUILDING 6

Partial Plan

C

A

E

Section

Partial elevation

Figure 7.24   Building 6.

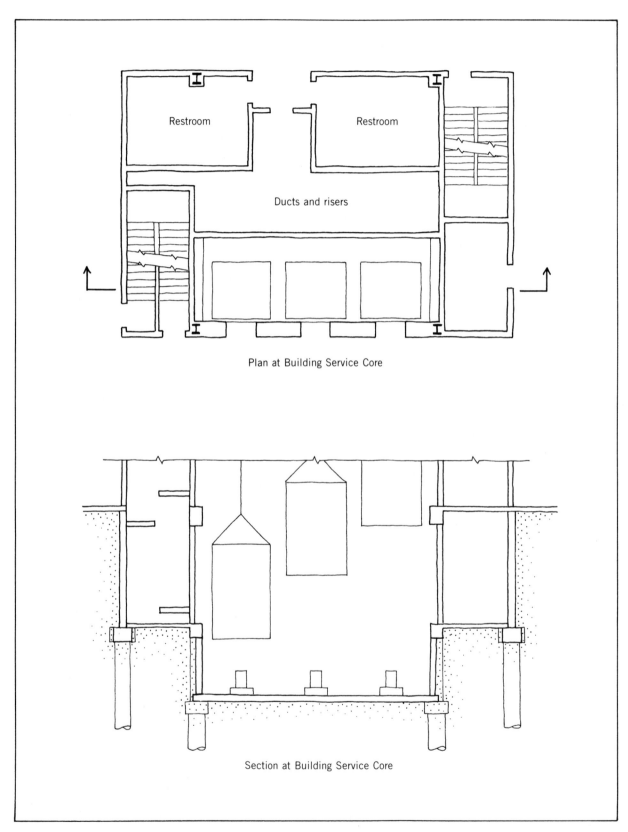

Plan at Building Service Core

Section at Building Service Core

Figure 7.25   Building 6: Core plan and section.

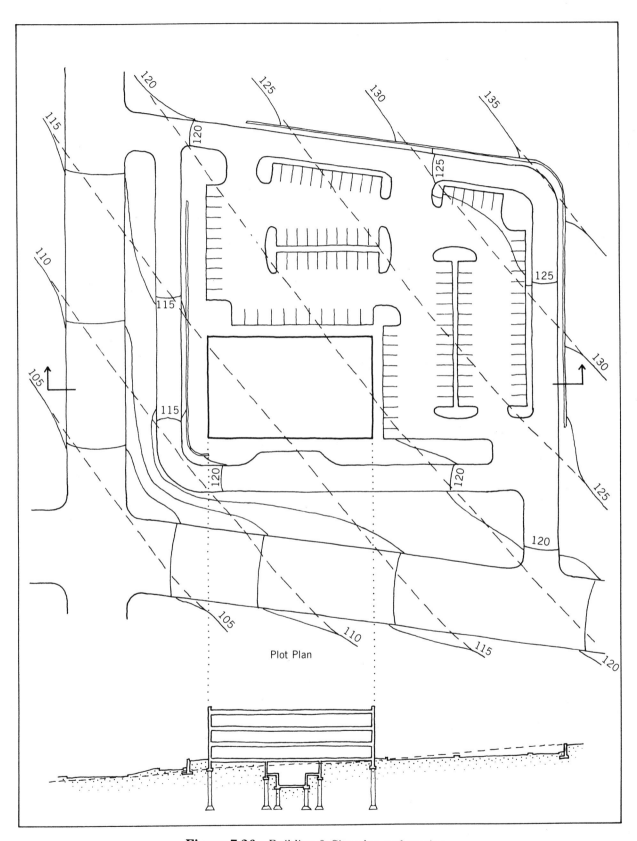

Figure 7.26 Building 6: Site plan and section.

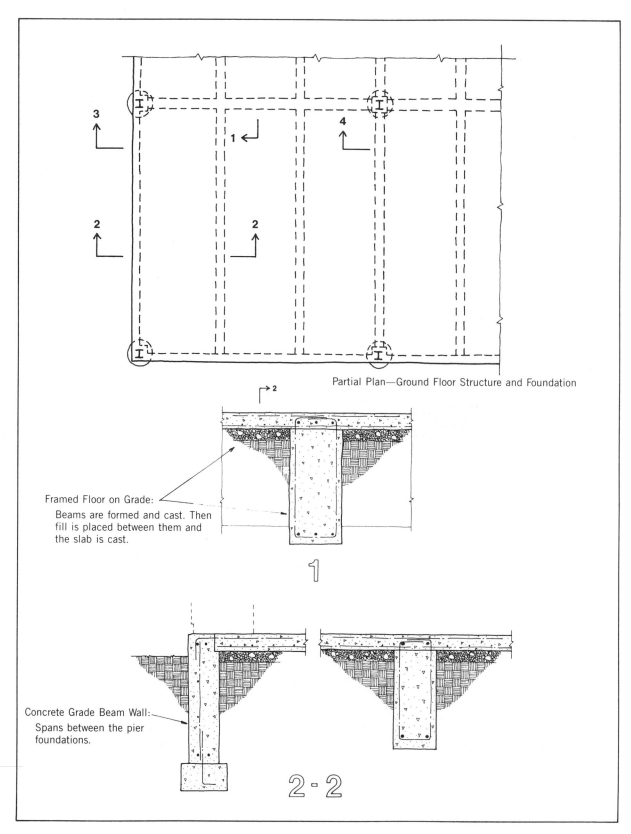

Partial Plan—Ground Floor Structure and Foundation

Framed Floor on Grade:

Beams are formed and cast. Then fill is placed between them and the slab is cast.

Concrete Grade Beam Wall:

Spans between the pier foundations.

Figure 7.27   Building 6: Ground floor and foundation plan and details.

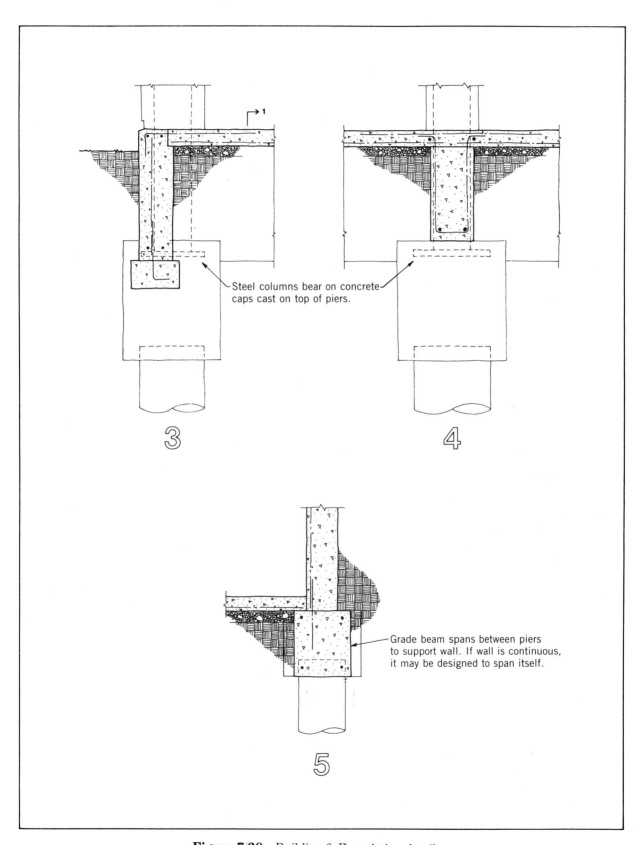

Figure 7.28   Building 6: Foundation details.

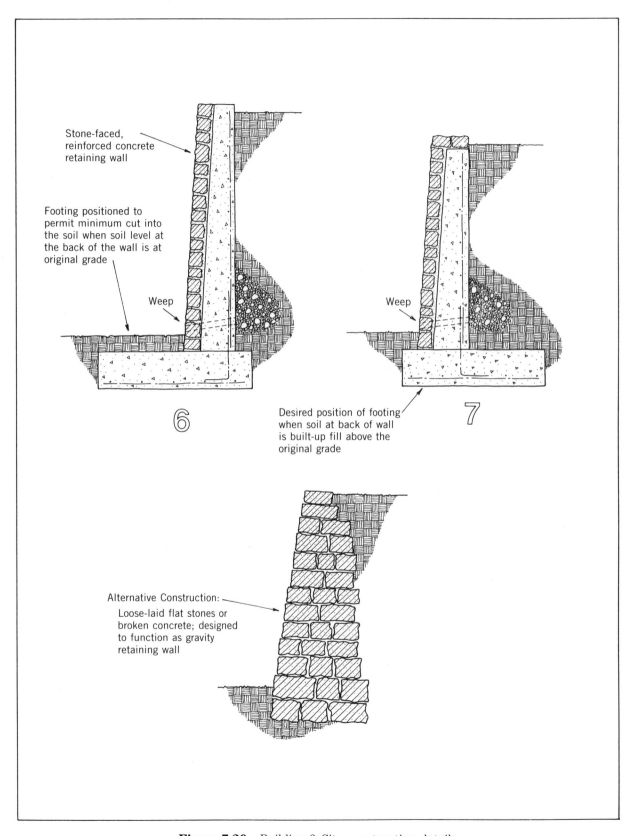

Stone-faced,
reinforced concrete
retaining wall

Footing positioned to
permit minimum cut into
the soil when soil level at
the back of the wall is at
original grade

Weep

6

Weep

7

Desired position of footing
when soil at back of wall
is built-up fill above the
original grade

Alternative Construction:

Loose-laid flat stones or
broken concrete; designed
to function as gravity
retaining wall

Figure 7.29   Building 6: Site construction details.

restrooms. The ground floor level is basically the lowest occupied level of the building. However, a partial basement level with equipment and storage is shown in the area of the core. The elevators also run down to this level, so an elevator pit extends even lower at this location.

The section in Figure 7.25 shows the general form of the construction of this lower level and the adjacent foundations. A more general foundation plan is shown in Figure 7.27 and a plan and details of the framed floor on grade for the general ground floor are shown on Figure 7.28.

The site plan and a site section are shown in Figure 7.26. The large site permits a generous space around the building and accommodates considerable surface parking. The site section indicates the cut and filled portions of the site and the general vertical position of the building with respect to the original grade. Because the ground level of the building is a considerable distance above the original grade, the concrete floor is developed as a framed structure, as shown in Figure 7.28.

Figure 7.27 shows a general foundation plan for the building and some details of the drilled pier foundations. Large grade beams between the piers are used to support the edges of the ground floor structure as well as the exterior walls. The core construction below the ground floor is also supported on the piers, as shown in Figures 7.25 and 7.27.

Figure 7.29 shows the general form of the large retaining walls used at the top and bottom edges of the sloped site. Details for these are slightly different, although the walls have basically a similar appearance. The uphill wall is built into a deep cut. Before the cut is made, a steel sheetpile wall is driven into the ground just slightly uphill of the retaining wall construction. The excavation of the cut then proceeds, with anchors used to brace the sheetpile wall as the cut gets deeper. When the cut is complete to the level of the bottom of the retaining wall footing, construction of the wall begins. When the concrete wall is completed, the steel wall may be either removed or left in place with a draining fill placed between it and the concrete wall.

The downhill wall, on the other hand, penetrates only a short distance into the original grade and serves basically to retain fill. This wall can therefore be built without the sheetpile wall. The positions of the footings for these two walls with respect to the wall are slightly different, reflecting the conditions for excavation and the proximity of the site boundaries.

7.7 BUILDING 7

Ten-Story Office Building
Concrete Frame, Metal Curtain Wall
Two-Level, Below-Grade, Plaza Structure

This building (see Figure 7.30) is not exactly high rise, but it is enough taller than building 6 to result in some narrowing of choices for the structure and some different planning concerns. It is unlikely that anything other than a steel or concrete frame would be used for this building. Shown here is a sitecast concrete structure with columns and a thick flat slab floor system.

The type of curtain wall used here forms a continuous skin outside the building structure. The columns at the building edge are thus almost freestanding inside the wall. For comparison, see the plan and wall form for Building 6, in which the columns are enclosed by the thick exterior wall at the floor level.

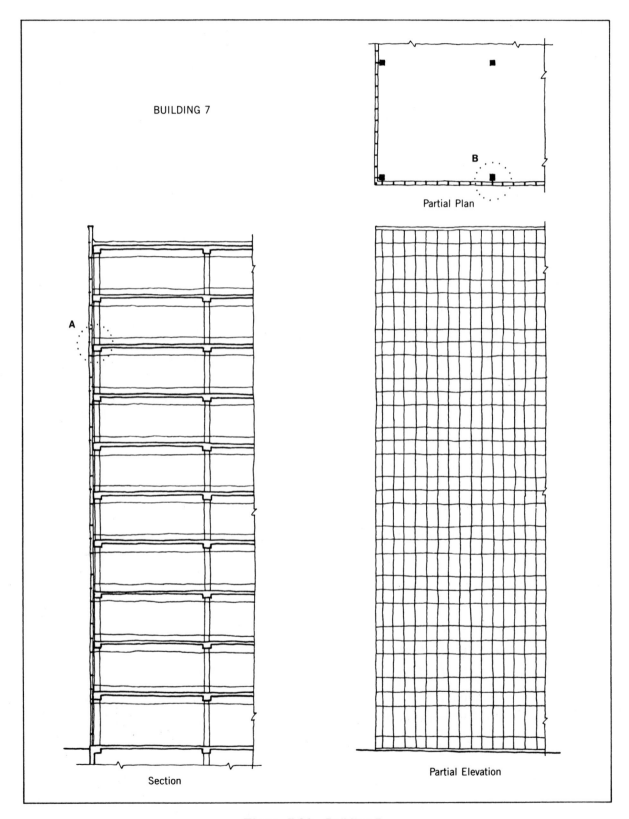

BUILDING 7

Partial Plan

Section

Partial Elevation

Figure 7.30   Building 7.

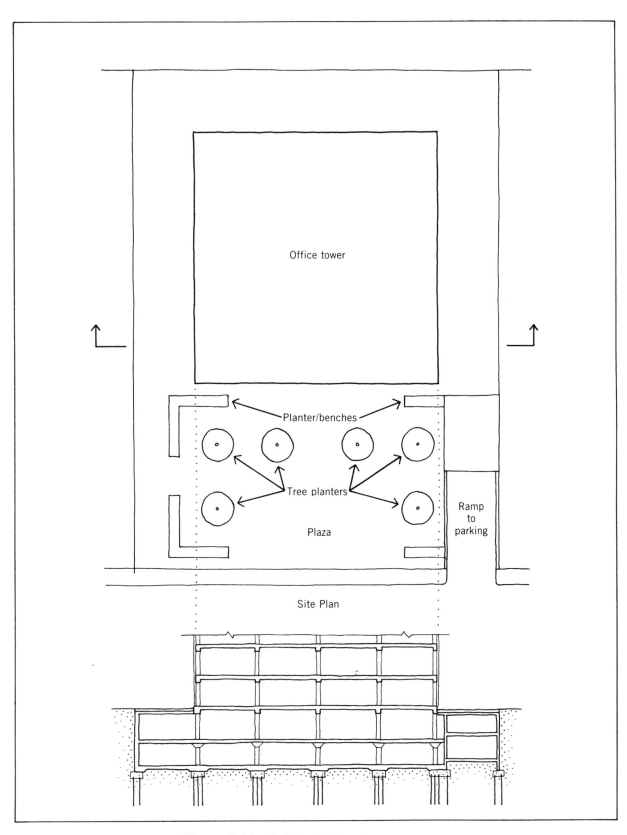

Figure 7.31   Building 7: Site plan and section.

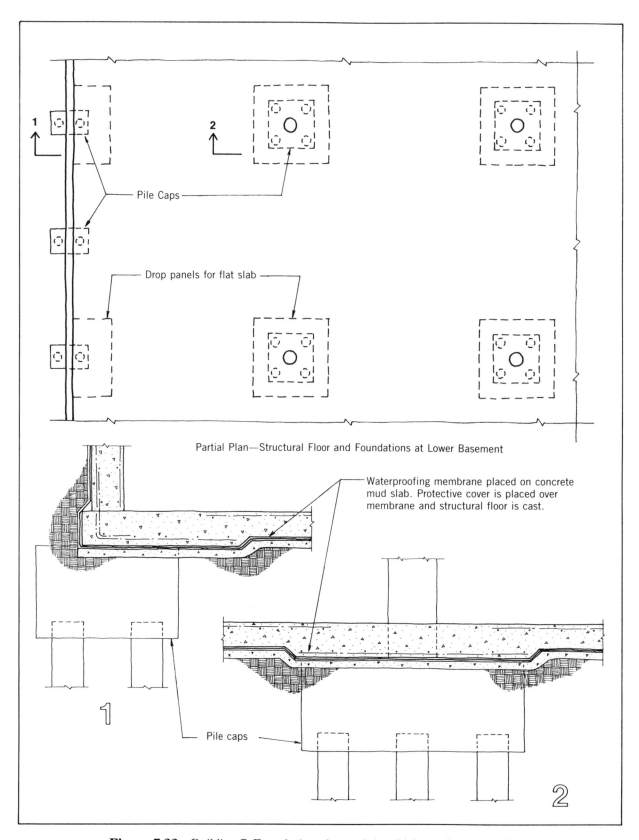

Figure 7.32   Building 7: Foundation plan and details, lower basement level.

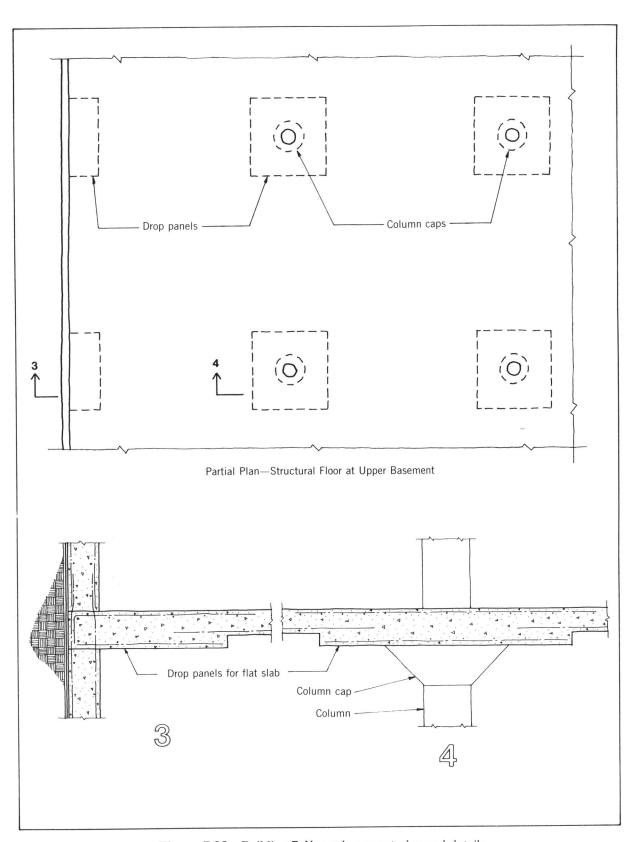

Partial Plan—Structural Floor at Upper Basement

Figure 7.33   Building 7: Upper basement plan and details.

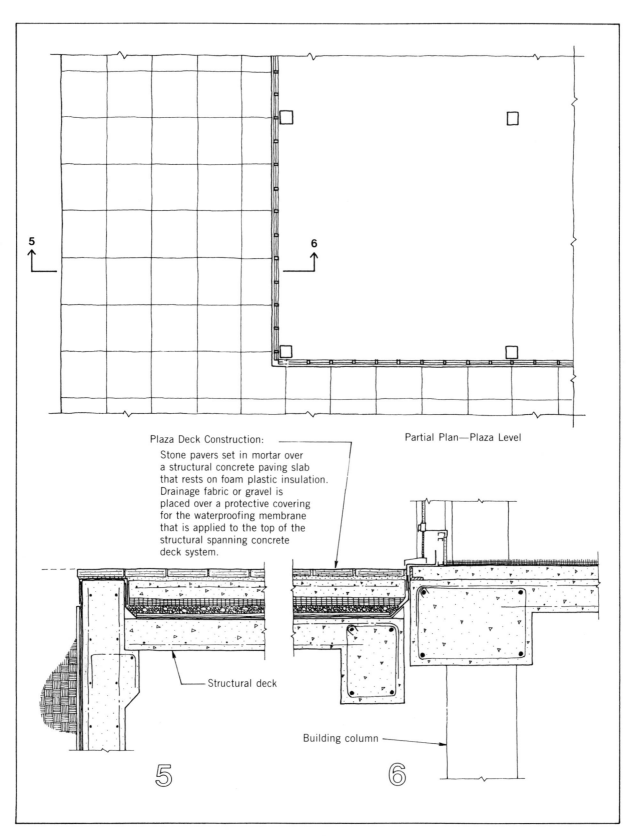

Plaza Deck Construction:

Stone pavers set in mortar over a structural concrete paving slab that rests on foam plastic insulation. Drainage fabric or gravel is placed over a protective covering for the waterproofing membrane that is applied to the top of the structural spanning concrete deck system.

Partial Plan—Plaza Level

Structural deck

Building column

Figure 7.34   Building 7: Ground floor plan and details, plaza level.

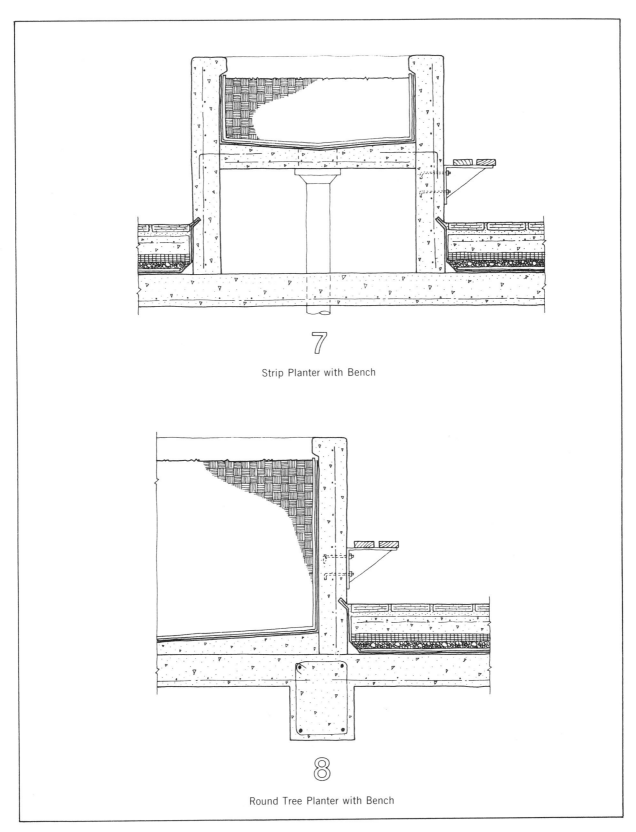

7

Strip Planter with Bench

8

Round Tree Planter with Bench

Figure 7.35 Building 7: Site construction details.

The curtain wall is formed using the "stick" system, in which vertical structural mullions are developed like a stud system to provide basic support for the rest of the wall. Windows continue in horizontal rows past the columns, so that the building exterior surface does not reveal the pattern of the column spacing, as it does in building 6.

Figure 7.31 shows a site plan and building/site section, which reveals that the office tower is surrounded by a plaza that is the roof of a two-story underground parking garage. Figure 7.32 shows the plan at the lowest level and also indicates the layout of the foundation system, consisting of piles. The piles are driven in clusters to support both the columns and the walls of the building.

Some details of the pile foundations are also shown in Figure 7.32. A problem with piles is the limited accuracy with which the locations of their tops can be controlled. Thus it is unreasonable to place a structure on a single pile, even though its vertical compression load capacity may be more than sufficient. The plans and details therefore show the use of a minimum of two piles for wall support and three for a single column.

Figure 7.33 shows some of the details for the construction of the basement walls and floors. It is assumed here that the lowest floor level is deep enough to have major water intrusion problems, so the bottoms of the walls and the lower-level floor are designed for a hydrostatic pressure assuming a groundwater level at some distance above the lower floor. The construction shown for the floor uses a so-called mud slab, which is placed on the soil and used as a base for installation of the waterproofing membrane and the structural slab.

The structural slab for the lower basement level must also be designed for the uplift force of the water pressure. This makes it function in a manner similar to that of the tower floor slabs, albeit in a reversed direction, and its design and construction may well be quite similar.

The upper basement floor is simply a spanning floor system, not unlike that for the office tower upper floors. Here also, a concrete flat slab or shallow waffle system may be used as an exposed structure.

Figure 7.34 shows a partial plan of a portion of the office tower and the plaza at the ground floor and plaza level, as well as some of the details for the plaza construction. There are various options for the form of the plaza paving and the waterproofing and draining of this structure. All options involve considerable total weight, so the supporting structure here will be quite heavily loaded.

Figure 7.35 shows some details of additional elements of the plaza-level development. Planters may be quite shallow for small plantings, but must be appropriately deep for large shrubs or trees.

7.8 BUILDING 8

Sports Arena, Longspan Roof
Steel Two-Way Truss System
Interior, Partly Below-Ground Concrete Structure

This is a medium-size sports arena, possibly big enough for a swim stadium or a basketball court (see Figure 7.36). Options for the structure here are strongly related to the desired building form. Functional planning requirements derive from the specific activities to be housed and from the seating, internal traffic, overhead clearance, and exit and entrance arrangements.

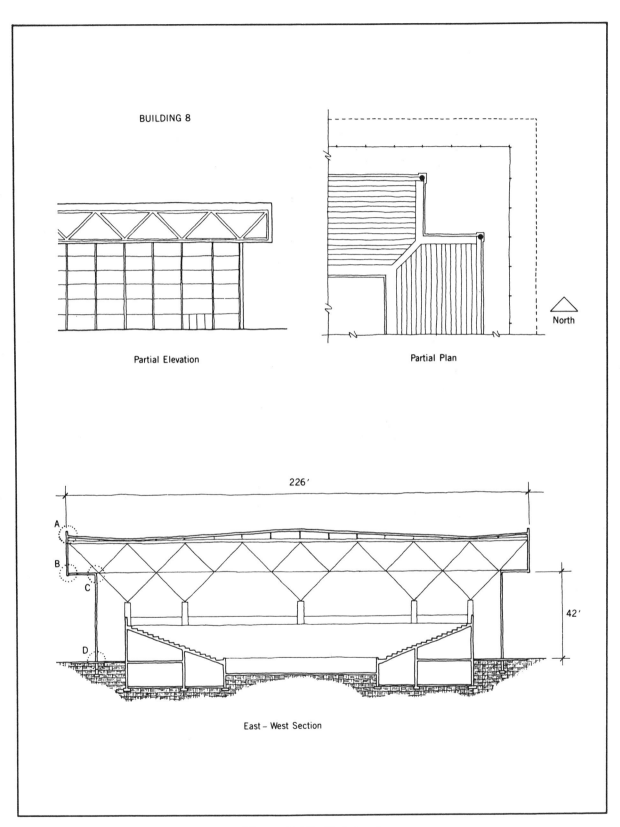

BUILDING 8

Partial Elevation

Partial Plan

North

226'

42'

East – West Section

Figure 7.36   Building 8.

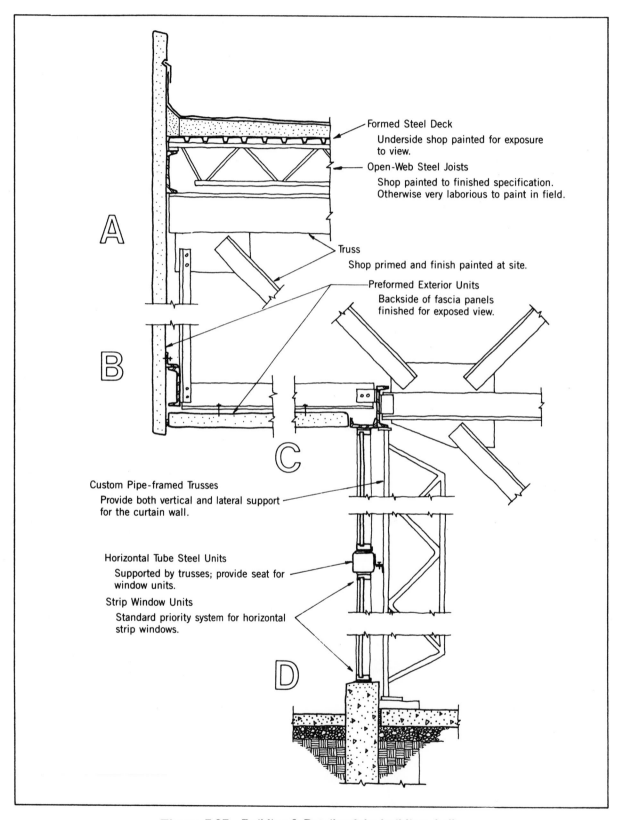

Formed Steel Deck
    Underside shop painted for exposure
    to view.

Open-Web Steel Joists
    Shop painted to finished specification.
    Otherwise very laborious to paint in field.

Truss
    Shop primed and finish painted at site.

Preformed Exterior Units
    Backside of fascia panels
    finished for exposed view.

Custom Pipe-framed Trusses
    Provide both vertical and lateral support
    for the curtain wall.

Horizontal Tube Steel Units
    Supported by trusses; provide seat for
    window units.

Strip Window Units
    Standard priority system for horizontal
    strip windows.

Figure 7.37   Building 8: Details of the building shell.

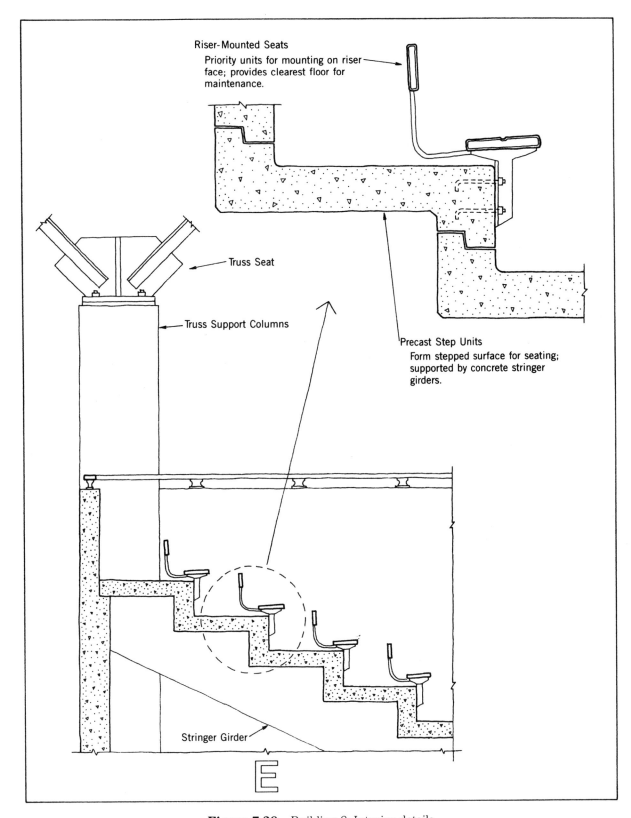

Figure 7.38   Building 8: Interior details.

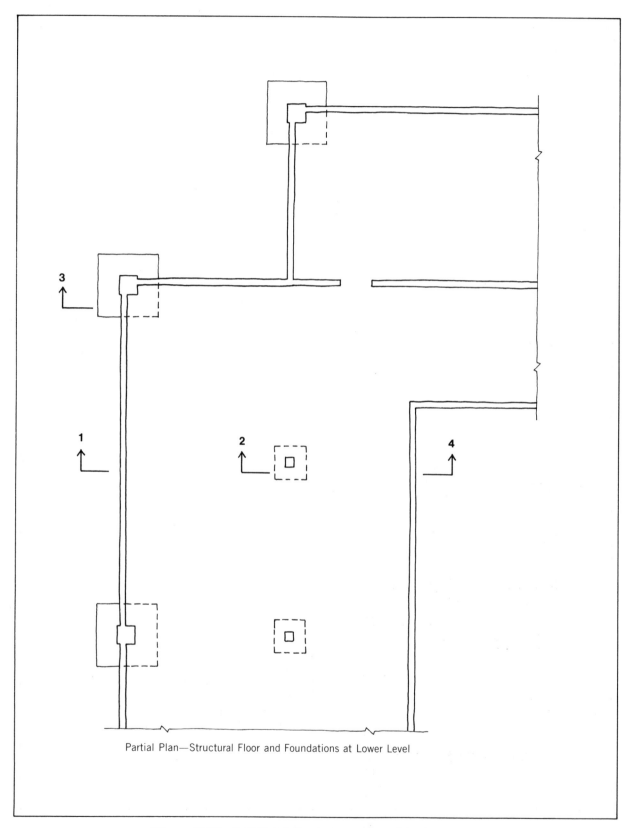

Partial Plan—Structural Floor and Foundations at Lower Level

Figure 7.39   Building 8: Basement and foundation details.

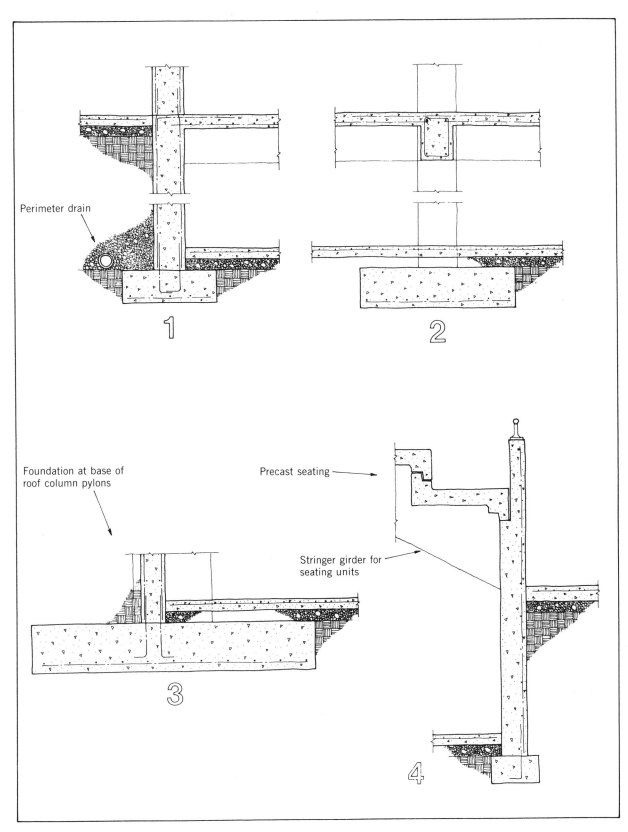

Figure 7.40 Building 8: Foundation details.

In spite of all the requirements, there is usually some room for consideration of a range of alternatives for the general building plan and overall building form. Choice of the system shown here relates to a commitment to a square plan and a flat roof profile. Other choices for the structure may permit more flexibility in the building form or also limit it. Selection of a dome, for example, would require a round plan, or at least one that could be developed under the hemispherical enclosure. Need for an oblong plan would limit the possibility for a two-way spanning structure.

The structure shown here uses a two-way spanning steel truss system (often called a space frame, although the name is quite ambiguous). Basic planning with this type of structure requires the use of a module that relates to the nodal points (joints) of the truss system. Locations of supports must relate to the nodal points, and any major concentrated loads should be applied at nodal points. The use of the nodal-point module may be extended to general planning of the building, as has been done in this example.

The details in Figure 7.37 show the general form of the roof and the tall glass curtain wall at the exterior. Both the interior, entry-area floor and the paved area around the exterior are developed as simple concrete slabs on grade.

The building shell, of course, exists only to house the building activities. In this case, an essentially independent construction is developed under the large overhead enclosure to form the seating and other required facilities. The support columns for the truss are incorporated into this construction, but the separation is otherwise complete and the planning could be totally independent. The "roof" of this structure constitutes the seating for the arena, which is shown in Figure 7.38.

A partial plan of the lower level and foundations is shown in Figure 7.39. Details of the foundations for the roof columns are shown in Figure 7.40. General construction of the lower structure uses exposed concrete basement walls and slab floors.

The details of the arena floor would depend on the particular forms of activity anticipated here. This may be a simple dirt floor for some events, but would be considerably more complicated for ice skating or swimming. This is mostly an independent consideration, although the use of the lower structure's spaces would also be related.

7.9 BUILDING 9

Open Pavilion
Wood Pole Frame and Timber Truss
Concrete Site Structure Seating

This is an open-air facility of medium span, consisting only of a large roof over a small amphitheater that is built into the ground (see Figure 7.41). Foundations and vertical supports for the roof structure are provided by wood poles with their bottom ends buried in the ground. The roof structure consists of a system of timber trusses supporting wood purlins and a wood plank roof deck. The roof surface is developed with a sheet metal roofing system with standing ribs.

Beneath the roof is a separate concrete structure that provides seating and defines the depressed amphitheater area in the center. Timber seats are bolted to the stepped concrete seating area. Steps on four sides lead down to the seating and the amphitheater. Ground surfaces are developed with pulverized bark and loose stone paving.

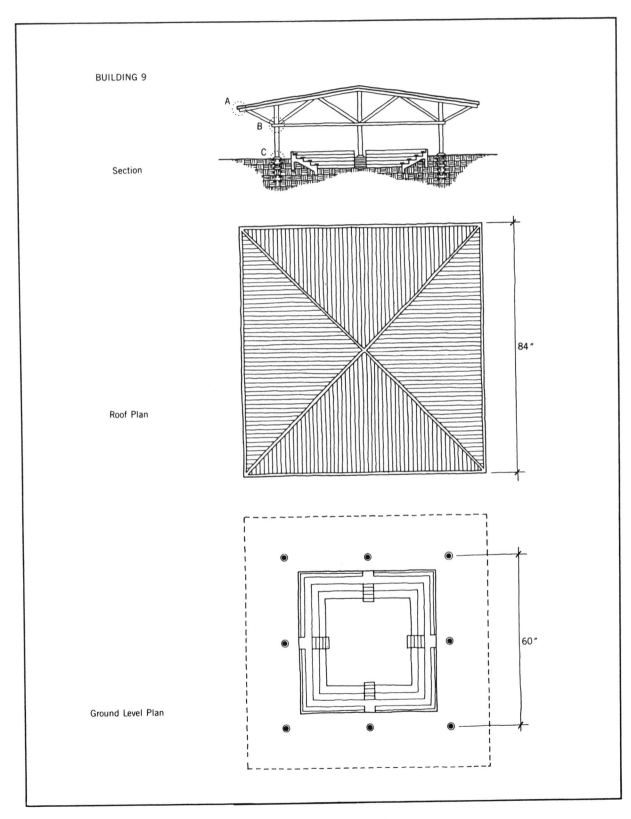

Figure 7.41   Building 9.

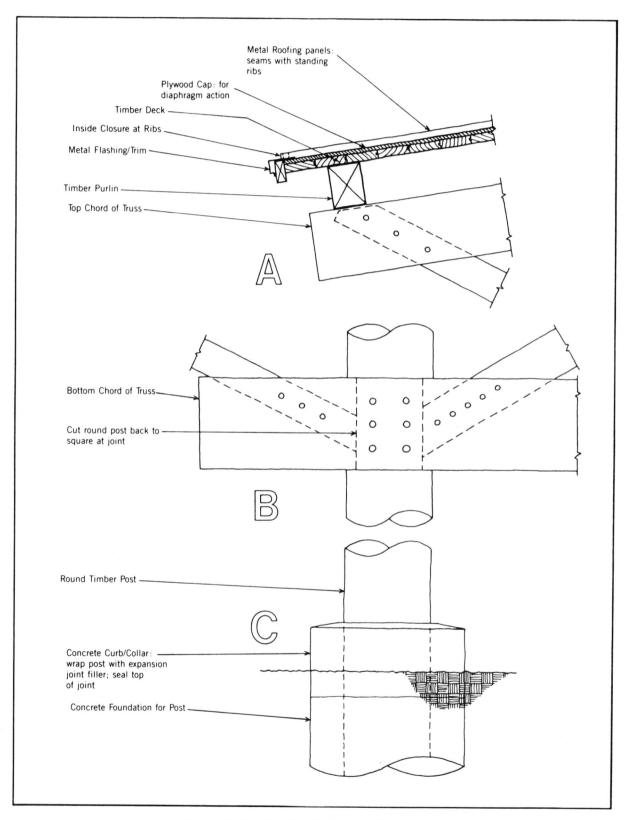

Metal Roofing panels:
seams with standing
ribs

Plywood Cap: for
diaphragm action

Timber Deck

Inside Closure at Ribs

Metal Flashing/Trim

Timber Purlin

Top Chord of Truss

A

Bottom Chord of Truss

Cut round post back to
square at joint

B

Round Timber Post

C

Concrete Curb/Collar:
wrap post with expansion
joint filler; seal top
of joint

Concrete Foundation for Post

Figure 7.42   Building 9: Construction details.

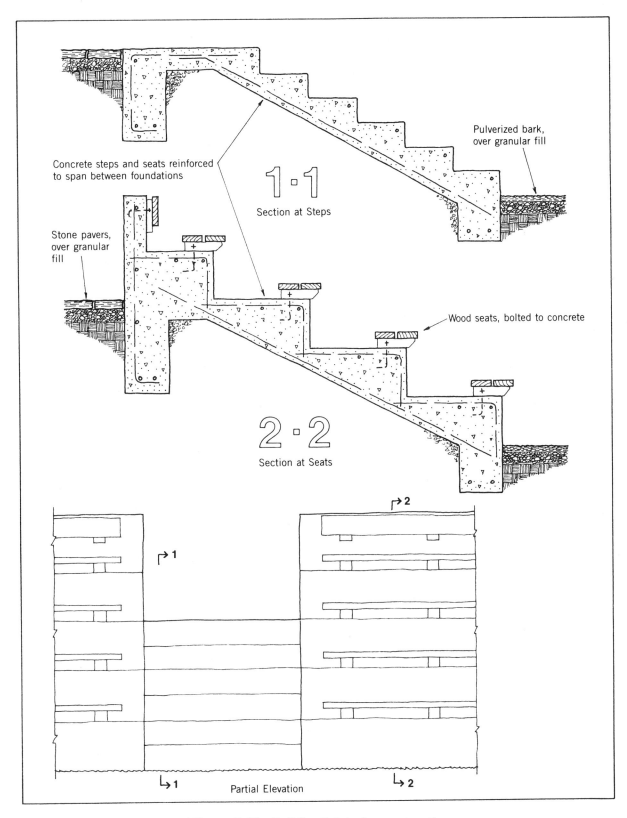

Concrete steps and seats reinforced to span between foundations

Pulverized bark, over granular fill

1·1

Section at Steps

Stone pavers, over granular fill

Wood seats, bolted to concrete

2·2

Section at Seats

Partial Elevation

Figure 7.43   Building 9: Interior construction.

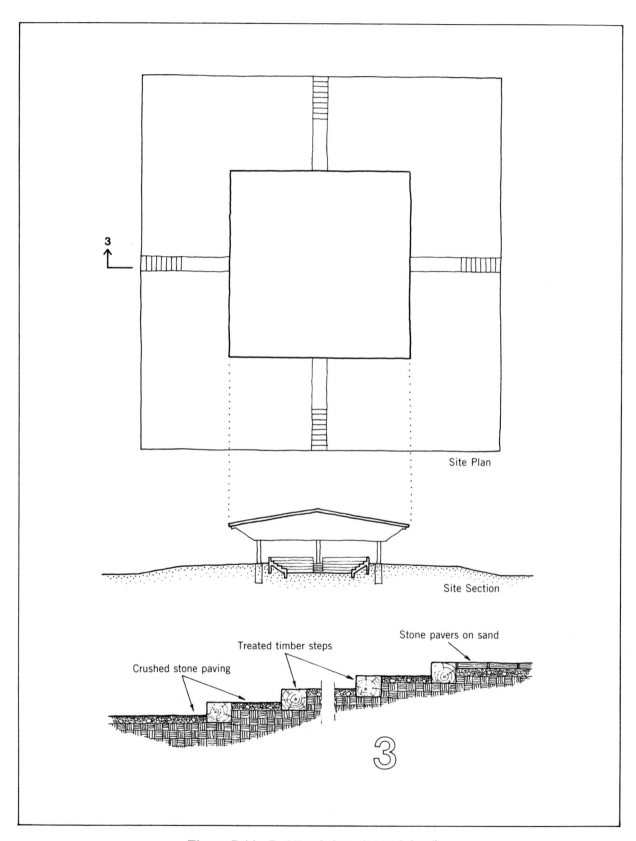

Site Plan

Site Section

Stone pavers on sand

Treated timber steps

Crushed stone paving

3

Figure 7.44   Building 9: Site plan and details.

This is indeed more of a site structure than a building in the usual sense. The roof is simply a large canopy, sheltering a seating system built into the ground. It is a somewhat primitive structure, with some level of roughness in the finished quality—especially the timber poles, which may be simple tree trunks with only approximately round sections, tapered profiles with changing diameters, and many knots, splits, and so on. The truss members and purlins may also be rough, but could be standard structural lumber.

The roof deck is visible only on the underside, and thus only this surface is of concern for appearance. The plank units may be solid sawn, but are increasingly likely to be laminated, with the viewed surface chosen for appearance. General details of the wood structure are shown in Figure 7.42.

There is some analogy here with the structure for building 8, although the different materials, smaller size, and lack of an enclosing wall create considerable differences. What *is* common is the development of the interior with the exposed underside of the roof and the exposed structure. Also, as with building 8, the activity being housed is quite independent from the roof structure in terms of construction.

The only interior construction in this case is that provided for the seating, which defines the form of the amphitheater. Some details for the seating are shown in Figure 7.43.

The general site plan and a site section are shown in Figure 7.44, indicating that the building is situated on a rise of ground and the site is generally pitched down away from the seating area. This should be sufficient to assure that the depressed amphitheater area does not become a pool in wet weather. If not, a drain field could be placed under the amphitheater with the drain lines running out and downhill to a leaching field.

Appendix

Soil Properties and Behaviors

The materials in this appendix present a general summary of concerns for the soils that constitute the surface and subsurface ground mass of most sites. This material has been abstracted from various references, but mostly from *Simplified Design of Building Foundations*.

Information about the materials that constitute the earth's surface is forthcoming from a number of sources. Persons and agencies involved in fields such as agriculture, landscaping, highway and airport paving, waterway and dam construction, and the basic earth sciences of geology, mineralogy, and hydrology have generated research and experience that is useful to those involved in general site development.

A.1 SOIL CONSIDERATIONS RELATED TO SITE DEVELOPMENT

Some of the fundamental properties and behaviors of soils related to concerns in site development are the following:

Soil strength: For bearing-type foundations a major concern is the soil resistance to vertical compression. Resistance to horizontal pressure and to sliding friction are also of concern for situations involving the lateral (horizontally directed) effects due to wind, earthquakes, or retained soil.

Dimensional stability: Soil volumes are subject to change, principally in stress or water content. These changes affect settlement of foundations and pavements, swelling or shrinking of graded surfaces, and general movements of site structures.

General relative stability: Frost actions, seismic shock, organic decomposition, and disturbance during site work can also produce changes in physical conditions of soils. The degree of sensitivity of soils to these actions is called their relative stability. Highly sensitive soils may require modification as part of the site development work.

Uniformity of soil materials: Soil masses typically occur in horizontally stratified layers. Individual layers vary in their composition and thickness. Conditions can also vary considerably at different locations on a site. A major early investigation that must precede any serious engineering design is that for the soil profiles and properties of individual soil components at the site. Depending on the site itself and the nature of the work proposed for the site, this investigation may need to proceed to considerable depth below grade.

Groundwater conditions: Various conditions, including local climate, proximity to large bodies of water, and the relative porosity (water penetration resistance) of soil layers, affect the presence of water in soils. Water conditions may affect soil stability, but also relate to soil drainage, excavation problems, need for irrigation, and so on.

Sustaining of plant growth: Where site development involves considerable new planting, the ability of surface soils to sustain plant growth and to respond to irrigation systems must be considered. Existing surface soils must often be modified or replaced to provide the necessary conditions.

The discussions that follow present various issues and relationships that affect these and other concerns for site development.

A.2 GROUND MATERIALS

Soil and rock: The two basic solid materials that constitute the earth's crust are soil and rock. At the extreme, the distinction between the two is clear: loose sand versus solid granite, for example. A precise division is somewhat more difficult, however, since some highly compressed soils may be quite hard, while some types of rock are quite soft or contain many fractures, making them relatively susceptible to disintegration. For practical use in engineering, soil is generally defined as any material consisting of discrete particles that are relatively easy to separate, while rock is any material that requires considerable brute force for its excavation.

Fill: In general, fill is material that has been deposited on the site, to build up the ground surface. Many naturally occurring soil deposits are of this nature, but for engineering purposes, the term fill is mostly used to describe *man-made fill* or other deposits of fairly recent origin. The issue of concern for fill is primarily its recent origin and the lack of stability that this represents. Continuing consolidation, decomposition, and other changes are likely. The uppermost soil materials on a site are likely to have the character of fill—man-made or otherwise.

Organic materials: Organic materials near the ground surface occur mostly as partially decayed plant materials. These are highly useful for sustaining new plant growth, but generally represent undesirable stability conditions for various engineering purposes. Organically rich surface soils (generally called *topsoil*) are a valuable resource for landscaping—and indeed may have to be imported to the site where they do not exist in sufficient amounts. For support of pavements, site structures, or building foundations, however, they are generally undesirable.

Investigation of site conditions is done partly to determine the general inventory of these and other existing materials, with a view toward the general management of the site materials for the intended site development. Critical concerns for this management are discussed in Section 5.1.

A.3 SOIL PROPERTIES AND IDENTIFICATION

The following material deals with the definition of various soil properties and their significance to the identification of soils and determination of soil behaviors.

Soil Composition

A typical soil mass is visualized as consisting of three parts, as shown in Figure A.1. The total soil volume is taken up partly by the solid particles and partly by the open spaces between the particles, called the *void*. The void space is typically filled by some combination of gas (usually air) and liquid (usually water). Several soil properties relate to this composition, including the following:

Soil weight: Most of the materials that constitute the solid particles in ordinary soils have a unit density that falls within a narrow range. Expressed as specific gravity (the ratio of the unit density to that of water), the range is from 2.6 to 2.75, or from about 160 to 170 lb/cu ft. Sands typically average about 2.65; clays about 2.70. Notable exceptions are soils containing large amounts of organic (mostly plant) materials. If the unit weight of a dry soil sample is determined, the amount of void can thus be closely approximated.

Void ratio: The amount of the soil void can be expressed as a percentage of the total volume. However, in engineering work, the void is usually expressed as the ratio of the volume of the void to that of the solid. Thus a soil with 40% void would be said to have a void ratio of 40/60 = 0.67. Either means of expression can be used for various computations of soil properties.

Porosity: The actual percentage of the void is expressed as the porosity of the soil, which in coarse-grained soils (sand and gravel) is generally an indication of the rate at which water flows through or drains from the soil. The actual water flow is determined by standard tests, however, and is described as the relative *permeability* of the soils.

Water content: The amount of water in a soil sample can be expressed in two ways: by the *water content* and by the *saturation*. They are defined as follows:

$$\text{Water content} = \frac{\text{weight of water}}{\text{weight of solids}} \times 100$$

$$\text{Saturation} = \frac{\text{volume of water}}{\text{volume of void}}$$

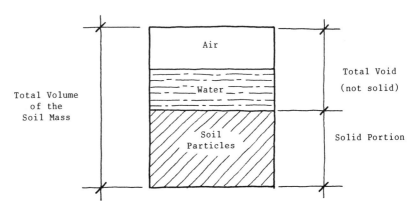

Figure A.1   Three-part soil composition.

Full saturation occurs when the void is totally filled with water. Oversaturation is possible when the water literally floats or suspends the solid particles, increasing the void. Muddy water is really very oversaturated soil.

Size and Gradation of Soil Particles

The size of the discrete particles that constitute the solids in a soil is significant with regard to the identification of the soil and the evaluation of many of its physical characteristics. Most soils have a range of particles of various sizes (expressed as the size *gradation*), so the full identification of the soil typically consists of determining the percentages of particles of particular size categories.

The two common means for measuring grain size are by sieve and by sedimentation. The sieve method consists of passing the pulverized dry soil sample through a series of sieves of increasingly smaller openings. The percentage of the total original sample that is retained on each sieve is recorded. The finest sieve is a No. 200, with openings of approximately 0.003 in.

A broad distinction is made between the total amount of solid particles that pass the No. 200 sieve and those retained on all the sieves. Those passing the No. 200 sieve are called the *fines* and the total retained is called the *coarse fraction.*

The fine-grained soil particles are further subjected to a *sedimentation test.* This consists of placing the dry soil in a sealed container with water, shaking the container, and measuring the rate of settlement of the soil particles. The coarser particles will settle in a few minutes; the finest may take several days.

Figure A.2 shows a graph that is used to record the grain size characteristics for soils. The horizontal scale uses a log scale, since the range of the grain size is quite large. Some common soil names, based on grain size, are given at the top of the graph. These are approximations, since some overlap occurs at the boundaries, particularly for the fine-grained materials. The distinction between sand and gravel is specifically established by the No. 4 sieve, although the actual materials that constitute the coarse fraction are often the same across the grain size range.

The curves shown on the graph in Figure A.2 are representative of some particularly characteristic soils, described as follows:

A *well-graded* soil consists of some significant percentages of a wide range of soil particle sizes.

A *uniform* soil has a major portion of the particles grouped in a small size range.

A *gap-graded* soil has a wide range of sizes, but with some concentrations of single sizes and small percentages over some ranges.

Shape of Soil Particles

The shape of soil particles is also significant for some soil properties. The three major classes of shape are bulky, flaky, and needlelike, the last being quite rare. Sand and gravel are typically bulky and are further distinguished as to the degree of roundness of the particle shape, ranging from angular to well rounded.

Bulky-grained soils are usually quite strong in resisting static loads, especially when the grain shape is quite angular, as opposed to well rounded. Unless a bulky-grained soil is well graded or contains some significant amount of fine-grained

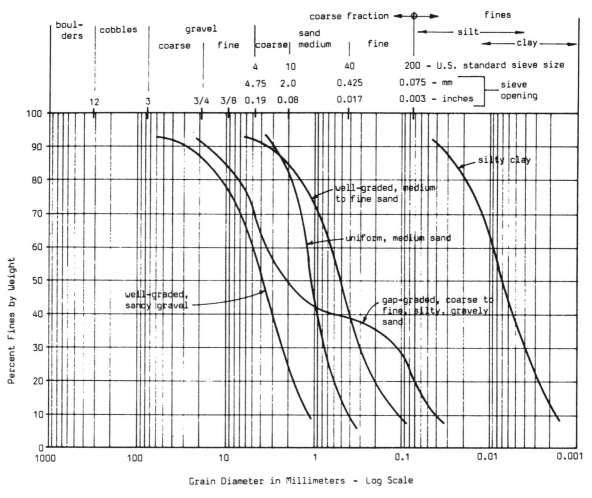

Figure A.2   Grain size range for typical soils. Reproduced from *Simplified Design of Building Foundations,* 2nd ed., with permission of the publishers, John Wiley & Sons.

material, however, it tends to be subject to displacement and consolidation (reduction in volume) due to shock or vibration.

Flaky-grained soils tend to be easily deformable or highly compressible, similar to the action of randomly thrown loose sheets of paper or dry leaves in a container. A relatively small percentage of flaky-grained particles can give the character of a flaky soil to an entire soil mass.

Effects of Water

Water has various effects on soils, depending on the proportions of water and on the particle size, shape, and chemical properties. A small amount of water tends to make sand particles stick together somewhat, generally aiding the excavation and handling of the sand. When saturated, however, most sands behave like highly viscous fluids, moving easily under stress due to gravity or other sources.

The effect of the variation of water content is generally most dramatic in fine-grained soils. These will change from rocklike solids when totally dry to virtual fluids when supersaturated.

Another water-related property is the relative ease with which water flows through or can be extracted from the soil mass—called the permeability. Coarse-grained soils tend to be rapid draining and highly permeable. Fine-grained soils tend to be non-draining or impervious, and may literally seal out flowing water.

Plasticity of Fine-Grained Soils

Table A.1 describes for fine-grained soils the Atterberg limits. These are the water content limits, or boundaries, between four stages of structural character of the soil. An important property for such soils is the *plasticity index,* which is the numeric difference between the liquid limit and the plastic limit.

A major physical distinction between clays and silts is the range of the plastic state, referred to as the relative plasticity of the soil. Clays typically have a considerable plastic range, while silts generally have practically none—going almost directly from the semisolid state to the liquid state. The plasticity chart, shown in Figure A.3, is used to classify clays and silts on the basis of two properties: liquid limit and plasticity. The diagonal line on the chart is the classification boundary between the two soil types.

Soil Structures

Soil structures may be classified in many ways. A major distinction is that made between soils considered *cohesive* and those considered *cohesionless.* Cohensionless soils consist primarily of sand and gravel with no significant bonding of the discrete soil particles. The addition of a small amount of fine-grained material will cause a cohensionless soil to form a weakly bonded mass when dry, but the bonding will virtually disappear with a small percentage of moisture. As the percentage of fine materials is increased, the soil mass becomes progressively more cohesive, tending to retain some defined shape right up to the fully saturated, liquid consistency.

The extreme cases of cohesive and cohesionless soils are personified by a pure clay and pure, or clean, sand, respectively. Typically, soils range between these extremes, making the extremes useful mostly for the defining of boundary conditions for classification.

TABLE A.1   Atterberg Limits for Water Content in Fine-Grained Soils

Description of Structural Character of the Soil Mass	Analogous Material and Behavior	Water Content Limit
Liquid	Thick soup; flows or is very easily deformed	
Plastic	Thick frosting or toothpaste; retains shape, but is easily deformed without cracking	Liquid limit: w_L — Magnitude of range is *plasticity index:* I_p
Semisolid	Cheddar cheese or hard caramel candy; takes permanent deformation but cracks	Plastic limit: w_P
Solid	Hard cookie; crumbles up if deformed	Shrinkage limit: w_S (Least volume attained upon drying out)

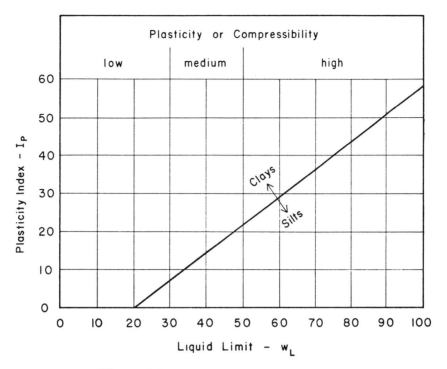

Figure A.3 Plasticity of fine-grained soils.

Sands

For a clean sand the structural nature of the soil mass will be largely determined by three properties: the particle shape (well rounded versus angular), the nature of size gradation (well graded, gap graded, or uniform), and the density or degree of *compaction* of the soil mass.

The density of a sand deposit is related to how closely the particles are fit together and is essentially measured by the void ratio or percentage. The actions of water, vibration, shock, or compressive force will tend to pack the particles into tighter (more dense) arrangements. Thus the same sand particles may produce strikingly different soil deposits as a result of density variation.

Table A.2 gives the range of density classifications that are commonly used in describing sand deposits, ranging from very loose to very dense. The general charac-

TABLE A.2 Average Properties of Cohesionless Soils

Relative Density	Blow Count, N (blows/ft)	Void Ratio, e	Simple Field Test with ½-in. Diameter Rod	Usable Bearing Strength (k/ft²)	(kPa)
Loose	<10	0.65–0.85	Easily pushed in by hand	0–1.0	0–50
Medium	10–30	0.35–0.65	Easily driven in by hammer	1.0–2.0	50–100
Dense	30–50	0.25–0.50	Driven in by repeated hammer blows	1.5–3.0	75–150
Very dense	>50	0.20–0.35	Barely penetrated by repeated hammer blows	2.5–4.0	125–200

ter of the deposit and the typical range of usable bearing strength are shown as they relate to density. As mentioned previously, however, the effective nature of the soil depends on additional considerations, principally the particle shape and the size gradation.

Clays and Silts

Many physical and chemical properties affect the structural character of clays. Major considerations are the particle size, the particle shape, and whether the particles are organic or inorganic. The amount of water in a clay deposit has a very significant effect on its structural nature, changing it from a rocklike material when dry to a viscous fluid when saturated.

The property of a clay corresponding to the density of sand is its consistency, varying from very soft to very hard. The general nature of clays and their typical usable bearing strengths as they relate to consistency are shown in Table A.3.

Another major property of fine-grained soils (clays and silts) is their relative *plasticity*. This was discussed previously in terms of the Atterberg limits, and the classification was made using the plasticity chart shown in Figure A.3.

Most fine-grained soils contain both silt and clay, and the predominant character of the soil is evaluated in terms of various measured properties, most significant of which is the plasticity index. Thus an identification as "silty" usually indicates a lack of plasticity (crumbly, friable, etc.), while that of "claylike" or "clayey" usually indicates some significant degree of plasticity (moldable without fracture, even when only partly wet).

Special Soil Structures

Various special soil structures are formed by actions that help produce the original soil deposit or work on the deposit after it is in place. Coarse-grained soils with a small percentage of fine-grained material may develop arched arrangements of the cemented coarse particles, resulting in a soil structure that is called *honeycombed.* Organic decomposition, electrolytic action, or other factors can cause soils consisting of mixtures of bulky and flaky particles to form highly voided soils that are called *flocculent.* The nature of formation of these soils is shown in Figure A.4. Water deposited silts and sands, such as those found at the bottom of dry streams or ponds, should be suspected of this condition if the tested void ratio is found to be quite high.

TABLE A.3   Average Properties of Cohesive Soils

Consistency	Unconfined Compressive Strength (k/ft^2)	Simple Field Test by Handling of an Undisturbed Sample	Usable Bearing Strength	
			k/ft^2	kPa
Very soft	<0.5	Oozes between fingers when squeezed	0	0
Soft	0.5–1.0	Easily molded by fingers	0.5–1.0	25–50
Medium	1.0–2.0	Molded by moderately hard squeezing	1.0–1.5	50–75
Stiff	2.0–3.0	Barely molded by strong squeezing	1.0–2.0	50–100
Very stiff	3.0–4.0	Barely dented by very hard squeezing	1.5–3.0	75–150
Hard	4.0 or more	Dented only with a sharp instrument	3.0+	150+

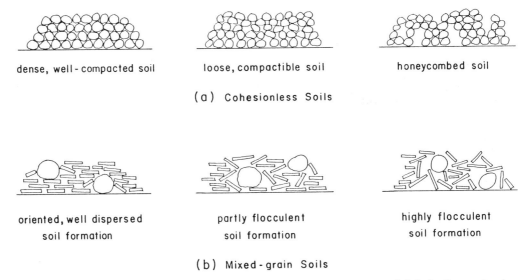

dense, well-compacted soil   loose, compactible soil   honeycombed soil

(a) Cohesionless Soils

oriented, well dispersed   partly flocculent   highly flocculent
soil formation   soil formation   soil formation

(b) Mixed-grain Soils

Figure A.4   Arrangements of particles in various soil structures: (a) In bulky-grained soils. (b) In soils with mixtures of bulky and flaky grains.

Honeycombed and flocculent soils may have considerable static strength and be quite adequate for foundation purposes as long as no unstabilizing effects are anticipated. A sudden, unnatural increase in the water content (such as that due to introduction of continuous irrigation) or some significant shock or vibration may disturb the fragile bonding, however, resulting in major consolidation of the soil deposit. This can produce major settlement of ground surfaces, pavements, or structures if the affected soil mass is extensive.

Structural Behavior of Soils

Behavior under stress is usually quite different for the two basic soil types, sand and clay. Sand has little resistance to compressive stress unless it is confined. Consider the difference in behavior of a handful of dry sand and sand rammed into a strong container. Clay, on the other hand, has some resistance to both compression and tension in an unconfined condition, all the way up to its liquid consistency. If a hard, dry clay is pulverized, however, it becomes similar to loose sand until some water is added.

In summary, the basic nature of structural behavior and the significant properties that affect it for the two extreme soil types are as follows:

Sand: Has little compression resistance without some confinement. Principal stress mechanism is shear resistance (interlocking particles grinding together). Important properties are penetration resistance to a driven object, unit density, grain shape, predominant grain size and size gradation, and a derived property called the *angle of internal friction.* Some reduction in capacity with high (saturated or supersaturated) water content.

Clay: Principal stress resistance in tension (cohesive tendency). Confinement generally of concern only in very soft, wet clays which may ooze or flow if compressed or relieved of confinement. Important properties are the tested unconfined compressive strength, liquid limit, plasticity index, and relative consistency (soft to hard).

These are, of course, the extreme limits or outer boundaries in terms of soil types. Most soils are neither pure clays nor clean sands and thus possess some characteristics of both of the basic extremes.

A.4 SOIL CLASSIFICATION AND IDENTIFICATION

Soil classification and identification must deal with a number of properties for precise categorization of a particular soil sample. Many systems exist and are used by various groups with different concerns. The three most widely used systems are the triangular textural system used by the U.S. Department of Agriculture; the AASHO system, named for its developer, the American Association of State Highway Officials; and the unified system, which is used in foundation engineering.

The unified system relates to properties that are of major concern in stress and deformation behaviors, excavation and dewatering problems, stability under loads, and other issues of concern to foundation designers. The triangular textural system relates to problems of erosion, water retention, ease of cultivation, and so on. The AASHO system relates primarily to the effectiveness of soils for use as base materials for pavements, both as natural deposits and as fill materials.

Figure A.5 shows the triangular textural system, which is given in graphic form and permits easy identification of the limits used to distinguish the named soil types. The

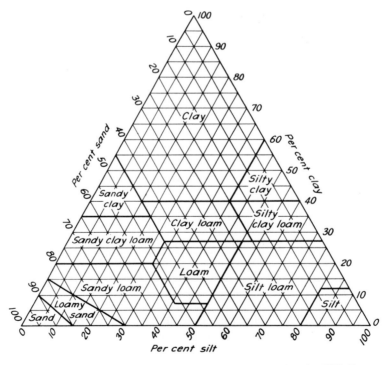

Figure A.5   Triangular textural classification chart used by the U.S. Department of Agriculture.

property used is strictly grain size percentage, which makes the identification somewhat approximate since there are protential overlaps between fine sand and silt and between silt and clay. For agricultural purposes this is of less concern than it may be for foundation engineering.

Use of the textural graph consists of finding the percentage of the three basic soil types in a sample and projecting the three edge points to an intersection point that falls into one of the named groups. For example, a soil having 46% sand (possibly including some gravel), 21% silt, and 33% clay would fall into the group called "sandy clay loam." However, it would be close to the border of "clay loam" and would in reality be somewhere between the two soil types in actual nature.

One use of the triangular graph is to observe the extent to which percentages of the various materials affect the essential nature of a soil. A sand, for example, must be relatively clean (free of fines) to be considered as such. With as much as 60% sand and only 40% clay, a soil is considered essentially a clay.

The AASHO system is shown in Table A.4. The three basic items of data used are the grain size analysis, the liquid limit, and the plasticity index, the latter two properties relating only to fine-grained soils. On the basis of this data the soil is located by group, and its general usefulness as a base for paving is rated.

The unified system is shown in Figure A.6. It consists of categorizing the soil into one of 15 groups, each identified by a two-letter symbol. As with the AASHO system, primary data used includes the grain size analysis, liquid limit, and plasticity index. It is thus not significantly superior to that system in terms of its data base, but it provides more distinct identification of soils pertaining to their general structural behavior.

Building codes and engineering handbooks often use some simplified system of grouping soil types for the purpose of regulating foundation design and construction.

TABLE A.4   American Association of State Highway Officials Classification of Soils and Soil Aggregate Mixtures—AASHO Designation M-145

General Classification[a]	Granular Materials (35% or Less Passing No. 200)							Silt-Clay Materials (More than 35% Passing No. 200)			
	A-1		A-3	A-2				A-4	A-5	A-6	A-7: A-7-5, A-7-6
Group Classification	A-1-a	A-1-b		A-2-4	A-2-5	A-2-6	A-2-7				
Sieve analysis percent passing:											
No. 10	50 max										
No. 40	30 max	50 max	51 min								
No. 200	15 max	25 max	10 max	35 max	35 max	35 max	35 max	36 min	36 min	36 min	36 min
Characteristics of fraction passing No. 40:											
Liquid limit				40 max	41 min	40 max	41 min	40 max	41 min	40 max	41 min
Plasticity index	6 max		N.P.[b]	10 max	10 max	11 min	11 min	10 max	10 max	11 min	11 min
Usual types of significant constituent materials	Stone fragments— gravel and sand		Fine sand	Silty or clayey gravel and sand				Silty soils		Clayey soils	
General rating as subgrade	Excellent to good							Fair to poor			

[a] Classification procedure: With required test data in mind, proceed from left to right in chart; correct group will be found by process of elimination. The first group from the left consistent with the test data is the correct classification. The A-7 group is subdivided into A-7-5 or A-7-6 depending on the plastic limit. For $wp > 30$, the classification is A-7-6; for $wp > 30$, A-7-5.

[b] N.P. denotes nonplastic.

Major Divisions			Group Symbols	Descriptive Names
Coarse-Grained Soils — More than 50% retained on No. 200 sieve	Gravels 50% or more of coarse fraction retained on No. 4 sieve	Clean Gravels	GW	Well-graded gravels and gravel-sand mixtures, little or no fines
			GP	Poorly graded gravels and gravel-sand mixtures, little or no fines
		Gravels with Fines	GM	Silty gravels, gravel-sand-silt mixtures
			GC	Clayey gravels, gravel-sand-clay mixtures
	Sands More than 50% of coarse fraction passes No. 4 sieve	Clean Sands	SW	Well-graded sands and gravelly sands, little or no fines
			SP	Poorly graded sands and gravelly sands, little or no fines
		Sands with Fines	SM	Silty sands, sand-silt mixtures
			SC	Clayey sands, sand-clay mixtures
Fine-Grained Soils — 50% or more passes No. 200 sieve	Silts and Clays Liquid limit 50% or less		ML	Inorganic silts, very fine sands, rock flour, silty or clayey fine sands
			CL	Inorganic clays of low to medium plasticity, gravelly clays, sandy clays, silty clays, lean clays
	Silts and Clays Liquid limit greater than 50%		OL	Organic silts and organic silty clays of low plasticity
			MH	Inorganic silts, micaceous or diatomaceous fine sands or silts, elastic silts
			CH	Inorganic clays of high plasticity, fat clays
			OH	Organic clays of medium to high plasticity
Highly Organic Soils			Pt	Peat, muck and other highly organic soils

Figure A.6 Unified system for soil classification, ASTM designation D-2487.5.

A.5 SPECIAL SOIL PROBLEMS

A great number of special soil situations can be of major concern in site and foundation construction. Some of these are predictable on the basis of regional climate and geological conditions. A few common problems of special concern are discussed in the following material.

Expansive Soils

In climates with long dry periods, fine-grained soils often shrink to a minimum volume, sometimes producing vertical cracking in the soil masses that extends to considerable depths. When significant rainfall occurs, two phenomena occur that can produce problems for structures. The first is the rapid seepage of water into lower soil strata through the vertical cracks. The second is the swelling of the ground mass as water is absorbed, which can produce considerable upward or sideways pressures on structures.

The soil swelling can produce major stresses in foundations, especially when it occurs nonuniformly, which is the general case because of ground coverage by paving, buildings, and landscaping. Local building codes usually have provisions for design with these soils in regions where they are common.

Collapsing Soils

Collapsing soils are in general soils with large voids. The collapse mechanism is essentially one of rapid consolidation (volume reduction) as whatever tends to maintain the soil structure with the large void condition is removed or altered. Very loose sands may display such behavior when they experience drastic changes in water content or are subjected to shock or vibration.

The most common causes of soil collapse, however, are those involving soil structures in which fine-grained materials achieve a bonding or molding of cellular voids. These soils may be quite strong when relatively dry, but the bonds may dissolve when the water content is significantly raised. Weakly bonded structures may also be collapsed by shock or simply by excessive compression.

The two ordinary methods of dealing with collapsing soils are to stabilize the soil by introducing materials to partly fill the void and substantially reduce the potential degree of collapse, or to use some means to cause the collapse to occur prior to construction. Infusion with bentonite slurry may be possible to reduce the void without collapse. Water saturation, vibration, or overloading of the soil with fill may be used to induce collapse.

A.6 SOIL MODIFICATION

In connection with site and building construction, modifications of existing soil deposits occur in a number of ways. The recontouring of the site, placing of the building on a site, covering the ground surface with large areas of paving, installation of extensive plantings, and development of a continuous irrigation system all represent major changes in the site environment.

Major changes, done in a few months or years, are equivalent to ones that may take thousands of years to occur by natural causes, and the sudden disruption of the equilibrium of the geological environment is likely to have an impact that results in readjustments that will eventually cause major changes in soil conditions, particularly those near the ground surface. The single *major* modification of site conditions, therefore, is the construction work.

In some cases, in order to achieve the construction work, to compensate for disruptions caused by construction activity, or to protect against future potential problems, it is necessary to make deliberate modifications of existing soils. The most common type of modification is that undertaken to stabilize or densify some soil deposit by compaction, proconsolidation, cementation, or other means. This type of modification is usually done either to improve bearing capacity and reduce settlements, or to protect against future failures in the form of surface subsidence, slippage of slopes, or erosion.

Another common modification relates to changes in the groundwater and moisture conditions. Recontouring for channeled drainage, covering of the ground with buildings or paving, and major irrigation or dewatering cause changes of this type.

Deliberate changes may consist of altering the soil to change its degree of permeability or its water-retaining properties. Fine-grained materials may be leached from a mixed-grain soil to make it more cohesionless and permeable. Conversely, fine materials may be introduced to make a coarse-grained soil less permeable or to produce a more stable soil mass.

Except for relatively simple compaction of surface materials, all soil modifications should be undertaken only with the advice of an experienced soils consultant.

Index